Matter:
Solids, Liquids, and Gases

GEMS® Teacher's Guide for Grades 1–3

by
Kevin Beals and Carolyn Willard

Skills
Observing • Comparing • Sorting • Categorizing • Classifying • Testing
Discussing • Communicating • Applying • Posing Questions
Critical Thinking • Making Explanations from Evidence • Evaluating Evidence

Concepts
Matter • Solids • Liquids • Gases • Properties

Themes
Matter • Stability • Structure • Scale

Mathematics Strands
Logic and Language • Measurement

Nature of Science and Mathematics
Scientific Community • Interdisciplinary • Real-Life Applications
Science and Technology

Time
Six or seven 45–60-minute sessions

Great Explorations in Math and Science
Lawrence Hall of Science
University of California at Berkeley

Lawrence Hall of Science,
University of California,
Berkeley, CA 94720-5200

Director: Elizabeth K. Stage

Cover Design, Internal Design, and Illustrations: Lisa Klofkorn
Photography: Rita Davies, Teena Staller

Director: Jacqueline Barber
Associate Director: Kimi Hosoume
Associate Director: Lincoln Bergman
Mathematics Curriculum Specialist:
Jaine Kopp
GEMS Network Director:
Carolyn Willard
GEMS Workshop Coordinator:
Laura Tucker
Staff Development Specialists:
Lynn Barakos, Katharine Barrett, Kevin Beals, Ellen Blinderman, John Erickson, Karen Ostlund
Distribution Coordinator:
Karen Milligan
Workshop Administrator: Terry Cort
Trial Test and Materials Manager:
Cheryl Webb
Financial Assistant: Vivian Kinkead
Director of Marketing and Promotion:
Steven Dunphy
Editor: Florence Stone
Principal Publications Coordinator:
Kay Fairwell
Art Director: Lisa Haderlie Baker
Senior Artists: Carol Bevilacqua,
Lisa Klofkorn
Staff Assistants: Rachel Abramson,
Brandon Hutchens, Phoenix Jieh, Yvette Mauricia, Ashley Morris, Shuang Pan

Contributing Authors:
Jacqueline Barber, Katharine Barrett,
Kevin Beals, Lincoln Bergman, Susan Brady,
Beverly Braxton, Mary Connolly,
Kevin Cuff, Linda De Lucchi, Gigi Dornfest,
Jean C. Echols, John Erickson, David Glaser,
Philip Gonsalves, Jan M. Goodman,
Alan Gould, Catherine Halversen,
Kimi Hosoume, Susan Jagoda,
Jaine Kopp, Linda Lipner, Larry Malone,
Rick MacPherson, Stephen Pompea,
Nicole Parizeau, Cary I. Sneider,
Craig Strang, Debra Sutter, Herbert Thier,
Jennifer Meux White, Carolyn Willard

Initial support for the origination and publication of the GEMS series was provided by the A.W. Mellon Foundation and the Carnegie Corporation of New York. Under a grant from the National Science Foundation, GEMS Leaders Workshops were held across the United States. GEMS has also received support from: the Employees Community Fund of Boeing California and the Boeing Corporation; the people at Chevron USA; the Crail-Johnson Foundation; the Hewlett Packard Company; the William K. Holt Foundation; Join Hands, the Health and Safety Educational Alliance; the McConnell Foundation; the McDonnell-Douglas Foundation and the McDonnell-Douglas Employee's Community Fund; the Microscopy Society of America (MSA); the NASA Office of Space Science Sun-Earth Connection Education Forum; the Shell Oil Company Foundation; and the University of California Office of the President. GEMS also gratefully acknowledges the early contribution of word-processing equipment from Apple Computer, Inc. This support does not imply responsibility for statements or views expressed in publications of the GEMS program. For further information on GEMS leadership opportunities, or to receive a publications catalog please contact GEMS. We welcome comments and suggestions.

ISBN 978-0-924886-92-8
Printed on recycled paper with soy-based inks.

Library of Congress Cataloging-in-Publication Data

Beals, Kevin.
  Matter : solids, liquids, and gases : grades 1/3 / by Kevin Beals and Carolyn Willard.
    p. cm.
  ISBN-13: 978-0-924886-92-8 (trade paper)
  1. Matter--Properties--Experiments--Juvenile literature.
I. Willard, Carolyn, 1947- II. Title.
  QC173.36.B43 2007
  530--dc22

                              2007021753

ACKNOWLEDGMENTS

A very special thanks to Rita Davies and her students at Oxford School in Berkeley for allowing us to pilot test these activities in their classroom and making many helpful suggestions to improve the unit. Rita took many photographs and her students beautifully grace this guide. Several other photographs come from the classes of teachers in Medford, Oregon. All the teachers who field tested this guide are listed in the "Reviewers" section at the back of this book. Their comments and suggestions contributed greatly to this GEMS guide.

We are very grateful for the scientific review of this unit by physicist Bruce Birkett. Bruce has reviewed a number of GEMS units and we are always appreciative of his awareness of and ability to help us walk that fine line between scientific accuracy and grade level appropriate clarity. Chemistry Professor Angelica Stacy of U.C. Berkeley also provided valuable input to this guide's early development. The highly creative Pat Lima originally developed the idea for Activity 2.

GEMS Director Jacqueline Barber's leadership spurred on this unit's development, helping ensure a "solid" conceptual framework and a strategic, research-based approach to student learning. GEMS Marketing Director Steven Dunphy supported our efforts to develop and complete this new GEMS guide in a multitude of ways. Thanks as well to Elaine Ratner for excellent editorial assistance.

We much appreciate the input of Chan Beals, a scientist and cousin of one of the authors of this guide. Chan contributed to the background section and provided expert scientific advice, particularly on ways to demonstrate that air has mass.

John Erickson of GEMS lent his scientific acuity and literary facility to revision of the background section and other related "matters."

Thanks all!

CONTENTS

TIME FRAME

The above guidelines are intended to give you a sense of how long the activities are likely to take. Please note that some field-test teachers needed some additional time, especially for Activities 1 and 2. The time you will need depends on variables such as students' ages and prior knowledge, their skills and abilities, your teaching style, and many other factors.

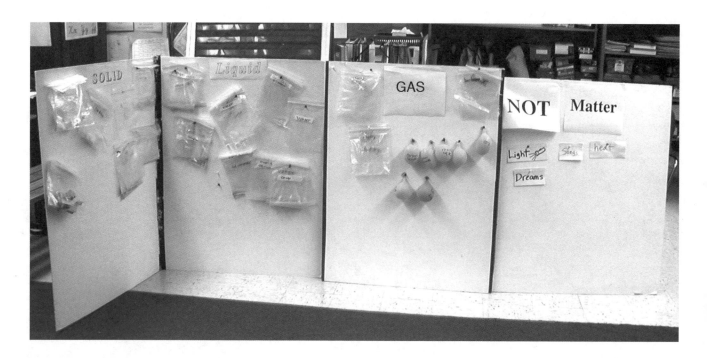

WHAT YOU NEED FOR THE WHOLE UNIT

The list of materials for this unit is long, but most items are fairly easy to obtain. Many of the materials can be donated by parents. A "Sample Letter to Parents" to use in asking for donations of materials is provided on page 9. Alternatively, a commercial kit containing everything you need, including harder-to- find items, such as spring scales, is available . Please see "Resources" on page 113, for additional tips on how to obtain and prepare materials.

The quantities below are based on a class size of 32 students. You may, of course, require different amounts for smaller or larger classes. This list gives you a concrete "shopping list" for the entire unit. Please also refer to the "What You Need" and "Getting Ready" sections for each individual activity, which contain more specific information about the materials needed for the class and for each group of students, and exactly how the materials are used.

Nonconsumables

- ❏ 1 plastic spoon
- ❏ 1 rock (big enough for the class to see)
- ❏ 1 piece of fabric at least a few inches square
- ❏ 3 transparent containers of different shapes (one with a lid)
- ❏ 1 cafeteria tray or plate
- ❏ 1 copy of Solids and Liquids signs
- ❏ about 150 pushpins (optionally, masking tape)
- ❏ 8 clear plastic bags to hold collections of solid and liquid items
- ❏ 32 clear plastic vials with tight-fitting lids
- ❏ 8 or more small glass beads or marbles
- ❏ 1 bag of cotton balls
- ❏ 8 or more small rocks or pebbles
- ❏ 2 boxes of wooden toothpicks
- ❏ 1 small box of metal paper clips (or 8 screws, nuts, bolts, washers, or coins)
- ❏ 8 small pieces of fabric
- ❏ about 150 sealable, clear plastic snack-size bags
- ❏ 4 cafeteria trays (to contain mess at some learning stations)
- ❏ 2 tweezers
- ❏ 2 plastic serrated knives
- ❏ 4 rolling pins (or mortars and pestles, or wooden hammers)
- ❏ 2 collecting devices for liquid (medicine droppers, turkey basters, etc.)
- ❏ 2 clear plastic cups
- ❏ 4 trash containers
- ❏ 6 wide-mouth plastic cups
- ❏ 4 dishtubs
- ❏ 1 one-teaspoon measuring spoon

- ❏ 1 one-tablespoon measuring spoon
- ❏ 1 stirrer
- ❏ 1 spring scale (calibrated, for teacher)
- ❏ 16 spring scales (for students)
- ❏ 16 paper clips (or binder clips)
- ❏ a large sign that says Gases
- ❏ 2 copies Making Gas procedure sign
- ❏ 1 vial or pill bottle
- ❏ 2 funnels
- ❏ 1 piece of wood
- ❏ 1 large, lightweight object (such as piece of Styrofoam)
- ❏ 1 small, heavy object (such as a fishing weight)
- ❏ 1 piece of plastic
- ❏ 1 small rock
- ❏ 1 piece of glass (not sharp)
- ❏ a pan and hot plate (or a shallow container or plate)
- ❏ 1 lightweight object (such as a pencil)
- ❏ 1 heavy object (such as a ring of keys)
- ❏ 1 very heavy object (such as a large rock)
- ❏ 8 hand magnifying lenses (or handheld microscopes)
- ❏ 1 copy of photo of helium balloon in a vacuum
- ❏ 4 film canisters with lids (or squeeze bottles)
- ❏ a few Styrofoam packing peanuts
- ❏ 1 air pump
- ❏ 1 hand-operated juicer
- ❏ *(optional)* 1 syringe without needle
- ❏ *(optional)* dissecting microscopes

Consumables

- ❏ water
- ❏ 1 16 oz. bottle of white vinegar
- ❏ 1 16 oz. box baking soda
- ❏ 4 squeeze bottles
- ❏ 4 graduated cylinders
- ❏ 12 plastic vials (no lids)
- ❏ 32 **Solids** student sheets (page 24)
- ❏ 32 **Liquids** student sheets (page 25)
- ❏ 32 **Solid or Liquid?** student sheets (page 54)
- ❏ 32 **What's the Matter?** student sheets (page 67)
- ❏ 32 **Gases** student sheets (page 86)
- ❏ 32 binders, clipboards, or folders (to keep journals in)
- ❏ 1 cup thick, clear liquid (such as shampoo or dishwashing detergent)
- ❏ 1 cup thick, opaque liquid (such as hair conditioner)
- ❏ 1 oz. red food coloring
- ❏ 1 oz. blue food coloring
- ❏ 20 sentence strips

- ❏ 1 roll of masking tape (or two sheets of address labels)
- ❏ 7 large plastic trash bags (to cover tables or desks)
- ❏ 16 plastic teaspoons
- ❏ several sheets aluminum foil
- ❏ 21 cotton balls
- ❏ 1 candle or chunk of wax
- ❏ 1 box of chalk
- ❏ 5 drops of food coloring (any color)
- ❏ 2 cups of dry cereal (that can be ground into a powder)
- ❏ 3 oranges
- ❏ a small amount of baking soda (in clear plastic cup)
- ❏ 1 cup of shaving cream
- ❏ $1/2$ tube of toothpaste
- ❏ 1 cup of sand
- ❏ 2 sealable plastic bags or other airtight containers
- ❏ 4 teaspoons of Borax powder
- ❏ 8 tablespoons of white glue
- ❏ 1 tablespoon vinegar (for making CO_2)
- ❏ 1 teaspoon baking soda (for making CO_2)
- ❏ about 20 balloons
- ❏ 2 identical balloons
- ❏ 1 tethered helium balloon
- ❏ 16 fold-top plastic sandwich bags
- ❏ a few drops of 4 scents, extracts, or perfumes
- ❏ 1 tablespoon peppermint extract (in closed bottle)
- ❏ about 5 lightweight plastic grocery bags
- ❏ scratch paper to make into paper fans
- ❏ 32 straws

General Supplies
- ❏ 32 pencils
- ❏ 6 black fine-tip permanent markers
- ❏ 2 black wide-tip felt markers
- ❏ chart paper
- ❏ several sheets of paper
- ❏ two 3 x 5 index cards
- ❏ paper towels
- ❏ 1 trash can
- ❏ 2 dustpans and brooms
- ❏ *(optional)* several rolls of masking tape (if pushpins not used for display)
- ❏ *(optional)* colored pencils or crayons
- ❏ *(optional)* 2–4 pairs of scissors
- ❏ *(optional)* 2 craft sticks
- ❏ *(optional)* a small container of glue (to seal lids on containers, if needed)
- ❏ *(optional)* 1 electric fan or hair dryer

INTRODUCTION

WARNING: THIS UNIT CAN LEAD TO A FASCINATION WITH SCIENCE.

Matter: Solids, Liquids and Gases invites students to figure out the physical world around them. As children, this is already their main pursuit! This unit capitalizes on that natural curiosity and helps young students learn to think critically and gather and apply evidence to expand their knowledge, just as scientists do.

Matter surrounds us, interacts with us, **is** us, all day every day! It's the stuff we can feel, hold, weigh, smell, see, touch, and taste. Chemistry is one major science discipline that is concerned with the study of matter, its composition, its properties, its behaviors, and how it changes. Given what matter is, its study is also central to physics and many other branches of science. In this guide we focus on the three basic states of matter—**solid, liquid,** and **gas.** The unit helps students generate definitions and understandings of the properties of solids, liquids, and gases, and apply these definitions and understandings to classification of "challenging substances."

Although these may seem like simple concepts, and although students may memorize various definitions, most students harbor persistent misconceptions regarding various aspects of these concepts. For this reason, students need to be given multiple opportunities to engage in diverse science experiences and wrestle with explanations in order to evolve more accurate and complete understandings over time. This unit can be an important, foundational part of that learning process.

Through the experiences and discussions in this unit, students learn that there are different types of matter. They discuss what is matter and what is not matter. They also gain experience in and understanding of science inquiry and the nature of science. The unit does not extend into investigation of the changing states of matter or phase change.

A Note on Energy and Matter

Physics and other sciences focus on the relationship between matter and energy. In the context of this guide and at this grade level, energy is not investigated. Energy is described solely by what it is not. It does not take up space or have mass—therefore it is not matter.

Foundations for Student Success

• **A foundation of science concepts.** An understanding of matter is key to the study of the sciences of chemistry, physics, geology, and astronomy. The ability to distinguish among solids, liquids, and gases is a conceptual gateway to more advanced understanding of matter, phase change, and the natural world. These concepts are also central components of the *National Science Education Standards, Benchmarks for Science Literacy,* state standards and district science curriculum frameworks. This unit stands solidly in line in with science standards. (see page 3)

• **A foundation of inquisitiveness.** Students are encouraged to think, to wonder, and to ask and attempt to answer questions. They are also encouraged to apply ideas and questioning strategies learned in class to their everyday surroundings. Research shows that student learning is deepened and becomes more lasting when students are confronted with substances that are challenging to classify or that demonstrate surprising or discrepant properties or behaviors. They do precisely that in this unit.

• **A foundation of positive attitude toward science.** When their classroom experiences cause students to recognize that their ideas and thinking are valued, they develop an attitude that they can do science. This is key to helping cultivate lifelong science learners. The inquiry-based, guided discovery approach to teaching can make even difficult and abstract concepts accessible to students, as it cultivates positive attitudes toward learning about and doing science.

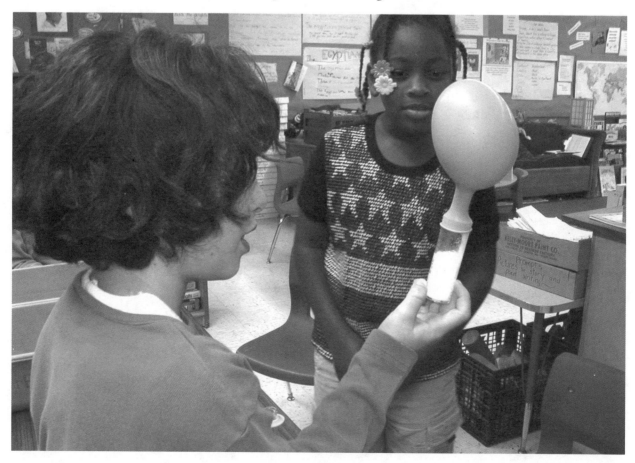

Matter
and
Relevant National and State Content Standards

(drawn from the *National Science Education Standards, Benchmarks for Science Literacy* and multiple state science standards)

This guide focuses on what its title states—*Matter: Solids, Liquids, and Gases*—and thus helps fulfill a fundamental aspect of Physical Science standards, as found in virtually all science standards, benchmarks, state and district frameworks. While these documents reflect some differences in phrasing, emphasis, and grade level expectations, they all prominently address building students' basic understanding of matter. This GEMS unit is in strong alignment with standards, including the following concepts:

- Materials can exist in different states—including solids, liquids, and gases. Solids, liquids, and gases have different properties and can be defined by and differentiated from each other by these properties.

  - Water can be a liquid or a solid or a gas and can go back and forth from one form to the other.

  - Air is a substance that surrounds us and takes up space. It is a mixture of gases.

- Objects have many observable properties, including size, weight, shape, color, texture, flexibility, temperature, and the ability to react with other substances. Those properties can be measured using tools, such as rulers, balances, and thermometers.

- Objects are made of one or more materials, such as paper, cloth, clay, wood, and metal. Objects can be described by the properties of the materials from which they are made, and those properties can be used to separate or sort a group of objects or materials.

- A lot of different materials can be made from the same basic materials.

- Most things are made of parts. Collections of pieces (powders, marbles, sugar cubes, or wooden blocks) may have properties that the individual pieces do not.

This unit's content and activities also help students meet expectations in other standards-related areas, especially relating to inquiry understandings and abilities.

How the Unit Addresses Misconceptions About Matter

Note: Please see pages 93 and 94 of the Background for the Teacher section for more information on student misconceptions about matter.

In order for students to build an understanding of matter, they often need to overcome certain logical but mistaken ideas they may have. There has been much research done on naïve or alternate conceptions that students (and others) may have about matter. The unit is designed to give students repeated opportunities to confront and revise their ideas by gathering physical evidence and discussing how it relates to concepts about matter. We provide suggestions throughout the guide for discussion strategies that teachers can use when they encounter some of these ideas among their students. In addition to issues related to solids, liquids, and challenging substances, the category of gases poses particular difficulties. Because gases are invisible, children may, for example, mistakenly think gases are not matter and do not have mass. Many other alternate ideas that students have are stubborn. While this unit can play an effective role in deepening understanding, unraveling these mistaken ideas often takes many experiences over considerable time.

Discussion and Evidence Are Crucial

Teachers and educational researchers recognize that for students to learn key concepts, they need time to reflect on their experiences, discuss differing viewpoints, raise questions, and consider evidence. Because this unit is about building concepts, discussion time is very important. Your students will already have many ideas about the substances they investigate. Some of these ideas will not hold up when examined critically and/ or when compared with evidence, and some will. Some student misconceptions may derive from lack of direct concrete experiences, others from explanations they've heard (including from teachers), and still others from lacks in logical thinking and/or incomplete thought processes.

Although many of the concepts in this unit could be quickly transmitted by a teacher and repeated by the students, full comprehension requires time to take root. It's important for the teacher to constantly assess student understanding and adjust the discussion accordingly. It's also wise to recognize that some of the ideas students bring up that may seem like digressions can lead to fruitful avenues of discussion and to further explorations. If your students have not had a lot of experience with group discussion, they may struggle with the process at first, but the ability to discuss ideas is a necessary life skill worthy of practice for its own sake, in addition to its learning impact.

In the context of class discussions, perhaps the most important objective is to help students gain experience in supporting their claims and arguments with **evidence** (measurements, data, observations) and to encourage their use of the language of logical scientific argumentation. Supporting one's ideas with evidence is central to science. Scientists base their explanations on evidence; they question, discuss, check, and debate each other's evidence and explanations. When new evidence warrants, scientists set aside a previous explanation for one that is better supported by the evidence.

This unit provides an outstanding platform for students to begin to acquire and hone such skills of argumentation, and to begin to better understand what distinguishes actual evidence—"the rock has mass because it moved the scale and takes up space"—from assumptions, inferences, or claims that are not based on any physical evidence at all.

Preparation and Materials

To conduct inquiries into matter, you need "stuff"! The list of materials for this unit is substantial, but so are the educational benefits! Most of the materials are inexpensive and accessible. And once the materials are gathered and prepared, future presentations of the unit will be much easier. Two important pieces of advice are:

1. Get volunteer helpers. Arrange in advance for parents or other volunteers to help you gather and prepare the materials. Certain pages of this teacher's guide can be used to delegate tasks to volunteers.

- Use the Sample Letter to Parents on page 9 to solicit donations of materials.

- Give the "What You Need" section for each Activity or the "What You Need for the Whole Unit" list to volunteers. Indicate on the list which materials you have on hand and what you still need. (It helps if you've already read the guide, so you understand how each item will be used.)

• Use the "Getting Ready" sections for each activity to communicate to volunteers the steps you need help with. In some activities the "Getting Ready" steps are divided into "Before the Day of the Activity" and "On the Day of the Activity."

2. Consider team-teaching the unit. Teachers sharing the preparation is a practical way to maximize the educational impact and minimize the time and energy spent by each teacher. Invite one or more other teachers to teach the unit concurrently. Alternatively, you could offer to prepare this unit for two (or more) classes, in exchange for your colleague's help in team-teaching something else.

The Learning Station Format

During part of almost every main activity in the unit, students work in pairs at hands-on learning stations. The use of stations allows for independent work and requires fewer materials than if you had to build a class set of every item. The "Getting Ready" sections give tips on how to set up the learning stations around the room and the guide has suggestions for managing students in this format. The stations are designed to be used in a flexible way, with student pairs moving to another station when they have completed one, rather than on a timed rotation. This allows for:

• Different lengths of time needed at different stations.

• Differing ability levels of students.

• The opportunity for students to pursue their own interests.

Five Main Activities

The five main activities in this unit are usually done in six or seven class sessions.

In *Activity 1: Solids and Liquids,* groups of four students explore collections of common materials. They identify what material each object is made of, then sort the objects into groups of their own choosing. The teacher leads a Secret Sort in which she sorts the objects into two groups (solids and liquids) and the students silently try to guess what the groups are. Each student suggestion is tested and evaluated to see if it is supported by the evidence. The teacher then reveals the categories of Solids and Liquids and defines them. A class display is created that will build throughout the unit. The teacher posts the new definitions, and the student teams use them to re-sort their objects into solids and liquids. They draw pictures of solids and liquids in their journals.

In *Activity 2: Collecting Solids and Liquids,* student pairs go to various stations around the room collecting objects, sort them, and place them in either the Solids or Liquids section of the class display. The class then discusses the placement of the items. Part of the discussion focuses on powders and granular substances, which many students classify as liquids because they pour. The students are challenged to add "true statements" to the posted class definitions of solids and liquids. **Note: This activity may need to be divided into two class sessions.**

In *Activity 3: Challenging Substances,* the students are introduced to hand signals for solid and for liquid, and use them as the teacher holds up objects. In pairs, students go to four new learning stations. Each station presents a substance that is challenging to categorize as solid or liquid: shaving cream, toothpaste, sand, and Glook. The students explore the substances then try to categorize them. Later, in group discussion, particular focus is placed on sand, to further "solidify" the concept that granular solids are indeed solid.

In *Activity 4: What's the Matter?* the students are introduced to the term *matter,* which is defined as anything in the Universe that has mass and takes up space. This definition, along with Matter and Not Matter signs, are added to the class display. In pairs, students use spring scales and test a variety of objects to see if they are matter. In a large-group discussion the class realizes that *all* solids and liquids are matter. The idea that air and other gases are matter is introduced briefly—but it is not fully presented until Activity 5. The students also discuss whether people are matter and conclude that yes, they are. Then the students are challenged to think of things that are *not* matter, and those the class agrees on are added to the class display.

In *Activity 5: Gases,* the students are introduced to the concept of gas. They are challenged to find evidence that air and other gases exist and are matter. During a peppermint extract demonstration, they learn that gases—unlike solids and liquids—are often invisible, but many can be detected by smell and they spread around in the air. Students go to

Students keep their own science journals throughout the unit. Research indicates that keeping a science journal or notebook over the course of a unit can contribute to lasting gains in student learning.

Suggestions for assessing student understanding of key concepts in the unit are provided on pages 108–112.

"gas stations," each of which involves an exploration of gases. Through teacher-led demonstrations they also see that gas has mass and takes up space, which makes it matter. They are taught a hand signal for gas, then challenged to classify a variety of objects as solids, liquids, or gases, or a combination. *Note: Activity 5 takes two class sessions.*

Application to the World Around Us

Learning about matter can be challenging (and exciting). Some students will catch on quickly, some may appear not to understand, and others will appear to understand but not comprehend fully. They can all benefit from repeated experiences and applications. Your students' grasp of this subject matter will be far greater if you encourage application of these new concepts of solid, liquid, gas, and matter while teaching other subjects throughout the day. This can be done by asking questions, such as:

- What am I filling this balloon with?
- Here's a strange substance—is it matter? Is it solid, liquid, or gas? What makes you think that?
- Look at the trees moving. What is moving them around?
- Is air solid, liquid, or gas? What makes you think that?
- Look at the lava in that film. Is it solid, liquid, or gas? What makes you think that?
- Turn on the lights—is the light matter? What makes you think that?
- What are you drinking? Is it matter? Is it solid, liquid, or gas? What makes you think it's a liquid?
- Is this hot chocolate powder solid, liquid, or gas? What makes you think that?
- What is that parachute floating on? Is it solid, liquid, or gas? What makes you think that? Is it matter?

By asking questions such as these, you can help cultivate a classroom atmosphere of collectively trying to figure out the surrounding world. With your help your class may "catch the fever" and carry inquisitiveness about the things around them to their homes, to everywhere they go—to the benefit of their lifelong learning!

Sample Letter to Parents

Dear Parents,

Your child's class will soon begin a unit called *Matter: Solids, Liquids, and Gases.* Students will observe, discuss, and classify many different items. In the process, they will learn important science skills and concepts and experience the actual work of science. This program was developed by the Great Explorations in Math and Science (GEMS) program at the Lawrence Hall of Science and tested in classrooms nationwide.

If you're able to volunteer an hour or two of your time during presentation of these active, hands-on lessons, it would be of great assistance! You can also help in other ways. In this unit, there are many materials that we need. Please see the attached list to see if you have any of the items to donate, or if you know of others who might have any of these materials.

We will need the materials by _____.

Thank you very much.

Overview

Activity 1 provides an introduction to the scientific processes students will use to investigate matter throughout the unit. It sets a tone of inquisitiveness and inquiry, as student groups try to understand common materials, discuss ideas, and make decisions based on evidence.

Students begin by working in groups of four to explore a variety of common materials. They observe the objects and sort them into groups of their own choosing. The teacher then gathers the class together and does a Secret Sort in which she sorts objects into two groups, but doesn't tell the students what the groups are. (The Secret Sort is by solids and liquids.) The students silently try to guess the sorting rule as items are added one by one. As students share their ideas, each idea is tested and evaluated to see if it is supported by the evidence.

Through a class discussion, the students are guided toward formal definitions of solids and liquids. The definitions are posted on a class display. Students will add to these definitions throughout the unit as they think of other "true statements" about solids and liquids, and as the class agrees that the available evidence supports such statements.

Groups of students apply the definitions as they re-sort their objects into groups of solids and liquids. They draw and label pictures of hard solids, soft solids, and liquids on student sheets. These student sheets are the opening pages of journals used throughout the unit.

One teacher said, "There was a buzz of positive, focused energy—all students engaged."

The concept of **evidence** is briefly introduced in this session, defined as **information we get from our senses that helps us understand things in the Universe.** While the definition is only touched on in this first, busy session, the students' understanding and use of evidence-based explanations will build in later sessions.

Learning Objectives for Activity 1

Students will:

- compare and sort objects according to their observable properties
- learn the definitions of solids and liquids
- use properties of solids and liquids to classify objects
- base their explanations on evidence
- follow directions, work cooperatively in groups, discuss, and record data

Notes on Adjusting Activity 1 for Your Situation

1. Adjusting for more experienced students. Teachers whose students have already had experience defining solids and liquids may want to skip the first sorting activity in Activity 1. For this option, begin the class session with the Secret Sort, then have small group of students sort the collections of items after the discussion.

2. Adjusting for younger or less experienced students. For most teachers, only one class session is needed for Activity 1, but some teachers of younger students may want to break it into two or even more segments. For students without much experience sorting and classifying, consider preceding the *Matter* unit with some practice sorts. Many primary mathematics curricula have suggestions for helping students practice sorting common items, such as buttons or bottle caps. The GEMS mathematics unit *Treasure Boxes,* for example, is an excellent way to deepen student understanding of sorting and classification, as well as graphing.

3. Streamlining this session to reduce time, preparation, and materials. As written, Activity 1 provides students with valuable direct experience sorting items and discussing their sorts in groups of four. Groups sort the items on their own. Later, after the class generates definitions of solids and liquids, they sort again. We recommend teaching the Activity as written if at all possible, but we know that time constraints are a reality. To significantly reduce the materials, preparation, and time needed, some teachers may decide not to have student groups sort bags of items. Instead, they may introduce the definitions of solids and liquids through the whole-class demonstration and Secret Sort only. For this option, only the materials listed under What You Need "For the class" are needed.

Materials are listed below for a class of 32 students. If you have a smaller class, you will need fewer items to sort. To decide exactly what you need, please see "Getting Ready," #3 (page 15).

■ What You Need

For the class
- ❏ 1 plastic spoon
- ❏ two 3 x 5 index cards, labeled #1 and #2
- ❏ 1 rock, any kind, big enough for the class to see
- ❏ 1 cotton ball
- ❏ 1 piece of fabric at least a few inches square
- ❏ 3 transparent containers of different shapes (one with a lid)
- ❏ 1 cafeteria tray or plate
- ❏ about 8–12 sentence strips
- ❏ a few pushpins or masking tape
- ❏ 1 wide-tip felt marker
- ❏ a space about 6–9 ft. wide on a bulletin board or wall for a display
  The display needs to be at a height students can reach and needs to stay up for the whole Matter *unit.*

For each student
- ❏ 1 binder, clipboard, or folder to keep journal pages together
- ❏ 1 copy of **Solids** student data sheet (page 24)
- ❏ 1 copy of **Liquids** student data sheet (page 25)

Materials to make solid/liquid collections
- ❏ 8 clear plastic bags★ to hold collections of solid and liquid items (see Notes about materials, on page 14)
- ❏ 32 clear plastic vials with tight-fitting lids
- ❏ 8 or more small glass beads or marbles
- ❏ 1 bag of cotton balls
- ❏ 8 or more small rocks or pebbles
- ❏ 1 box of wooden toothpicks, any kind
- ❏ 1 small box of metal paper clips (or 8 screws, nuts, bolts, washers, or coins)
- ❏ 8 small pieces of fabric
- ❏ about a cup of a thick, clear liquid like shampoo or dishwashing detergent
- ❏ about a cup of a thick, opaque liquid like hair conditioner
- ❏ 1 oz. of red food coloring
- ❏ 1 oz. of blue food coloring
- ❏ water
- ❏ *(optional)* a small container of glue to seal lids on containers, if needed

*Notes about materials

Clear plastic gallon-size freezer bags are ideal to hold the sets of items. You could also use clear plastic supermarket produce bags. Another option is to organize the sets of items on cafeteria trays..

You may want to substitute other items for the solids and liquids listed above. Each item should be made of just one material so it can be classified as either solid or liquid. Select liquids that will not spoil, so they can be stored and used again with future classes. If vials to hold the liquids are not available, other containers that will not leak or break, such as small plastic water bottles, will work. See the "Getting Ready" section below for more information about how the sets of items will be used.

■ Getting Ready

Before the Day of the Activity

1. **Decide if you need to pre-teach sorting.** During the first class session, groups of students will sort collections of objects into groups. If your students have not had experience sorting, you may want to take time before starting the unit to practice sorting with other common objects. Please see page 12 for ideas on easy and accessible sorting activities.

2. **Find space for the class display.** You and your students will add to a class display throughout the *Matter* unit. The display needs to be six to nine feet wide, and at a height students can reach. The display can be on a bulletin board or on butcher paper. Some teachers use part of the whiteboard or even a window.

   In Activity 2, the students attach small plastic bags with samples of solids and liquids to the display. If it's not possible for students to stick pushpins into the display, they can attach their bags with masking tape.

   Only a few items are added to the display during Activity 1, but by the end of Activity 5, the display will look something like this:

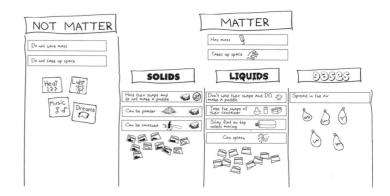

3. **Prepare bags of solid and liquid items.** For Activity 1, you will need a bag of about 10 solid and liquid items for each group of four students. The preparation directions below are based on a class size of 32, with eight groups of four students. If possible, teachers of younger students may want to have a bag for each pair of students, rather than each group of four.

For each group, prepare four vials (or other sealable containers) of different liquids. Fill each vial about half full. If the vials are not well sealed, keep them upright and hand them out to the groups separately rather than putting them in the bags. (Optionally, you could try using glue to better seal the lids.)

The bags of items take time to assemble, but they can be stored and reused with future classes. Parents may be able to donate much of what you need. (See Sample Letter to Parents, page 9.) See the inside front cover of this guide for information on obtaining a commercial GEMS materials kit. Whether you use the commercial kit or assemble your own, we recommend that you ask a parent or other volunteer to put the bags of items together for you.

   a. **Put six different solid items in a bag for each group:**
      - 1 small glass bead or marble
      - 1 cotton ball
      - 1 small rock or pebble
      - wooden toothpicks
      - metal paper clips (or a screw, nut, bolt, washer, or coin)
      - 1 small piece of fabric
      - *(optional)* additional solid items

   b. **Put four vials of liquid in each bag with the solid items:**
      - 1 vial of water with a few drops of red food coloring
      - 1 vial of water with a few drops of blue food coloring
      - 1 vial of a thick, clear liquid like shampoo or dishwashing liquid
      - 1 vial of a thick, opaque liquid like hair conditioner

On the Day of the Activity

1. **Write each of the following statements on a sentence strip in bold letters.** (You may need to tape sentence strips together to make longer ones.) These will be used on the class display, but **don't post them yet.** Ideally they should each also have an illustration representing the statement:

    • Hold their shape and do not turn into a puddle

    • Take the shape of their container

    • Stay flat on top, unless moving

    • Don't hold their shape, and do make a puddle

PLEASE NOTE: *We have provided three pages you can duplicate for signs (Solids, Liquids, Gases) on the last three pages of this guide, or you can make your own.*

2. **Make two large Solids and Liquids signs for the class display.**

> **SOLIDS**

> **LIQUIDS**

3. **Gather items for your whole-class demonstrations:**
    ___1 plastic spoon
    ___2 index cards labeled #1 and #2
    ___the 4 sentence strips you prepared earlier
    ___the Solids and Liquids signs you made
    ___1 rock
    ___1 cotton ball
    ___1 piece of fabric
    ___1 small transparent container filled with water
    ___2 empty transparent containers of different shapes, 1 with a lid
    ___1 cafeteria tray or plate onto which you can pour a little water
    ___2 vials of thick liquids (You can borrow these from the student bags if needed.)
    ___*(optional)* a few other objects

4. **Copy the two student data sheets, Solids and Liquids** (pages 24–25), for each student. These student sheets will be used again in later class sessions and will become part of student journals. Decide whether to use folders, binders, or clipboards for keeping the journals together.

5. **Plan groups.** Decide how to organize the students into groups of four.

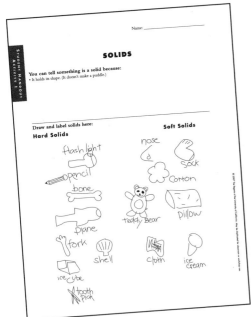

■ Observing Collections of Objects

1. **What is in the Universe?** Tell the class they're going to be studying the stuff that everything in the Universe is made of! Ask them to name some things in the Universe. [Accept all responses, since everything is.] Conclude by saying that everything in the whole world and in all of outer space is in the Universe.

2. **Free exploration of objects.** Tell the students their first job will be to explore a group of things or objects in the Universe and figure out what they are made of. Hold up a plastic spoon and ask what it is [a spoon], then ask what it is made of [plastic]. Hold up another item and ask what it is made of.

3. **Hold up one of the bags and say that each group will get a bag of objects.** Say they can take the objects out of their bag, and everyone in their group should look carefully at them, touch them, and talk about them, but they should not open the vials.

4. **Divide the class into groups of four.** Emphasize the importance of working cooperatively and sharing materials. Give each group a bag of objects and let them begin observing and discussing the objects. If the students focus on the liquids, encourage them to also explore some of the other items.

■ Sorting the Objects

1. **Regain attention of class.** After groups have had a few minutes to freely explore the objects, regain the attention of the whole class.

2. **Introduce the sorting activity.** Explain that scientists sometimes try to understand what's in the Universe by putting types of objects into groups. Tell the students they are going to be scientists and sort the objects in their bag into groups. Emphasize that there is no one right way to sort the objects.

   a. Ask for one or two examples of ways they could sort the objects. [For example, smooth versus rough objects, or objects that make noise when tapped against the table versus ones that do not.]

   b. Tell them everything in a group should be alike in some way. They can sort in any way they decide, as long as their group agrees on a reason for their sort. Also say that they can make two, three, or more groups of objects. Remind them to let everyone in their group help and participate.

If students name only astronomical objects, remind them that things like the chairs they're sitting on and the pen in your hand are also in the Universe.

Young students sometimes have a hard time distinguishing between objects and the materials they are made of.

c. Since only the liquids are kept in containers, students may suggest sorting according to what is in a container and what is not in a container. Explain that this is a possible way to group, but in this activity the containers are just holders, and students should focus on what is *inside* the containers.

3. **Have students begin sorting.** Circulate and challenge those who have sorted their objects to explain their reasoning. Ask early finishers to re-sort their objects in a different way. When all groups have finished sorting their objects in at least one way, have them replace the objects in their bags.

■ Introduce the Secret Sort

1. **Gather the class away from the bags of materials,** where they can hold a discussion and see your demonstration. Having students sit in a circle on a carpet is ideal. (They will go back to their bags of items at the end of this session.)

2. **Briefly discuss their sorts.** Ask a few students to describe how their group sorted the objects. "Why did you sort them the way you did?" Keep this sharing discussion very brief so there will be sufficient time for the Secret Sort and solids/liquids discussion.

3. **Explain the Secret Sort game.** Tell the class you're going to sort some objects into two groups, but you're not going to tell them why you're sorting them that way. It's their job to try to figure out what the secret rule you have in mind is as you sort.

Don't use the Solids and Liquids signs yet!

4. **Set the index cards labeled #1 and #2 on the carpet a foot or two apart.** Say all objects that you'll put in Group #1 will have something that's the same about them—something in common. All objects in Group #2 will also have something that's the same about them, but they will all be different in some way from the objects in Group #1.

5. **Tell the class the challenging part of the game is that they need to be silent.**

■ Play the Secret Sort Game

1. Dramatize your sorting process, and have students make predictions silently.

a. Hold up the rock. Look at it, poke it with your finger, shake it, set it on your hand, and lightly toss it in your hand. Set it near Card #1.

b. Hold up the container of water. Shake it. Take off the lid and poke your finger into the water. Silently feel the wetness on your finger in an exaggerated manner. Pour a small amount of water on your hand and look at it, then toss it in the air. Set the container near Card #2.

c. Take out another object and silently examine it in a similar manner.

- Hold the object above Group #1 and tell the students to silently raise their hands if they think it belongs in this group according to your Secret Sort.

  - Hold it above Group #2 and ask them to raise their hands if they think it belongs there according to your Secret Sort.

  - If the object is solid, place it in Group #1; if liquid, place it in Group #2.

d. Do the same with the other objects. As most students are starting to catch on, you can add to the fun by pretending to put a solid object in the liquids group, then finally placing it in the solids group and vice versa. Tell the class you think they have probably guessed your secret rule, but don't confirm it yet.

2. Ask, "What is the same about everything in Group #1?" Before you reveal your secret rule, call on a few students for ideas, and test their ideas.

a. For example, if a student says Group #1 is made up of *"things that are hard,"* touch each object in the group that is hard, poking each one for effect. Then poke a soft object. Acknowledge that many objects in the group are hard, but not all.

b. If a suggestion *does* apply to every object in the group, test each item and point this out.

c. If students come up with the term **solid,** ask what they mean by it. Using their definition, treat this idea the same as the others, pointing out that all the evidence seems to support their definition of solid.

3. Reveal the "secret" of Group #1. Place the Solids sign in front of Group #1.

The hand raising is meant to encourage all students to participate actively.

Instead of having students raise their hands, you may want to have students hold a hand close to their chests and give a thumbs-up or thumbs-down signal. This encourages students to vote their own opinion, rather than their neighbor's.

■ Define Solids

1. **Ask the students for their ideas about solids.** If no one has brought the term up yet, ask if they've heard of solids. Ask, "What makes an object a solid?" If they bring up an erroneous idea, probe to understand their thinking, and gently provide evidence that counters it, either now or during later discussions.

2. **Explain that scientists define a *solid* as something that holds its shape.**

   a. Ask, "When the rock is set on this tray (or plate), will it turn into a puddle?"

   b. Place the rock on the tray, and point out that it does not turn into a puddle. It holds its shape.

   c. Test this definition with a few more solid objects from your group. Be sure to include an object that is soft, to illustrate that **soft objects can be solids if they hold their shape.**

   d. Crumple up a piece of paper into a ball, and point out that you can change a solid object's shape, but *by itself,* it holds its shape.

   e. Set the sentence strip "Hold their shape and do not turn into a puddle" below the "Solids" sign.

■ Define Liquids

1. **Ask, "What is the same about everything in Group #2?"** Reveal that your rule was that every object in Group #2 is a *liquid.*

2. **Place the Liquids sign in front of Group #2.** Ask, "Who has heard of this term?" "What makes something a liquid?" Accept a few responses.

3. **Explain that scientists define a *liquid* as something that does not hold its shape.** Demonstrate.

   a. Hold up the container of water. Point out that the shape of the water is the same shape as the container it is in.

   b. Ask what shape the students think the water will be when it is poured out of the container onto the tray or plate. "Do you think it will hold its shape or make a puddle?"

Simply correcting a child tends to be a quick way to temporarily give students a correct answer. For deeper and lasting understanding it is better to expose the students to evidence that forces them to re-evaluate and self correct their ideas. In addition, this strategy models and teaches the scientific process of evaluating ideas based on evidence.

c. Pour a small amount of water onto the tray. Point out that it **does not** hold its shape; it **does** make a puddle.

d. Set the sentence strip "Don't hold their shape, and do make a puddle" below the Liquids sign.

e. Point out that liquids can be different from each other in many ways—such as in color, in whether or not you can see through them, or in whether they are thick or thin—but all liquids have in common that they do *not* hold their shape. They are not solids.

4. Add two more true statements about liquids.

a. Carefully pour the water from its container to a different-shaped open transparent container. Do this again with the third container, the one with a lid. Add the sentence strip "Take the shape of their container" below the Liquids sign.

b. Put the lid on the container. Point out that the top of the liquid looks flat. Tilt the container and ask what the top of the liquid looks like. [It's always flat once the motion stops.]

c. Add the sentence strip "Stay flat on top, unless moving" below the Liquids sign.

d. Challenge your students to come up with more "true statements" about solids and liquids. Test each one and if the class (and you) agree, write it on a sentence strip and add it to the others. Tell the class (and remind them periodically) that they can come up with more true statements throughout the unit.

Some students may bring up gases. If so, tell them gases are a third group, and the class with be learning more about gases later.

If they come up with a statement that is true about all the solids or liquids in the display, but you know is not true of some other solids or liquids not present, be sure to provide them with this evidence now or later.

Many teachers teach their students that solids can be defined as "objects you can't stick your finger through," and liquids as "things you can pour." There are problems with both of these statements. First, while there are certainly many solids you can't stick your finger through, such as rocks, there are also many that you can, such as paper or cotton. Second, although it's true that liquids do pour and it's fine to add this to your list as a "true statement" about liquids, by itself it doesn't work well as a definition or to distinguish liquids from gases or solids. That's because gases also can be poured, even though we may not always be able to see that happen, and tiny solids, such as sand, can also be poured. The key definitions included in this guide—that solids hold their shape and don't make a puddle, and liquids don't hold their shape and do make a puddle—are both accurate and accessible for young students.

■ Re-Sorting Their Objects and Recording in Journals

Adjusting for different writing abilities:

1. Teachers of younger students may not want to include writing in this first class session. In that case, sometime before Activity 2, provide time for groups to review the definitions you've discussed, sort the objects again, and record on the two data sheets.

2. For older or more experienced students, you might add a writing assignment after the Solids and Liquids sheets are completed. Have students pick one solid from the objects they sorted, draw it on a separate piece of paper, and describe why they classified it as a solid. Have them do the same with a liquid. These pages can be added to their journals.

Don't worry if students don't fully understand now what the word evidence *means. They will have many more opportunities later in the unit to deepen their understanding.*

You may want to start a "Word Wall." *As students request help with spelling, write the words on the wall on a large piece of chart paper for the whole class to use.*

1. **Tell the students they will get to re-sort their bag of objects into solids and liquids (though some may have already done this).** Everyone should use her senses to find *evidence* that the objects are solid or liquid.

2. **What is *evidence*?** Tell the students if we can see or feel that something holds its shape—that is *evidence* that it is a solid. Evidence is information we get from our senses. Evidence helps us figure out and explain things in science.

3. **Encourage students to use their senses to find evidence about the objects in their bags.** Ask, "Are the objects solids or liquids?" "How do you know?" Say that you'll put the Solids and Liquids signs and sentence strips up on the display.

4. **Hold up a Solids and a Liquids student sheet.** Point out where the students will draw pictures of solids and liquids. Review why a soft solid, such as a cotton ball or piece of fabric, should be classified as a solid. Show students where they will draw soft and hard solids on the **Solids** sheet. Say that after they draw all the solids and liquids from their collection, they can draw others from the room, or from memory. Also ask them to label each drawing (if they can).

5. **Pass out the two student sheets and a folder or binder to each student.** Help them set up their journals.

6. **Have the groups return to their bags of objects.** As the students begin drawing, assemble the signs and sentence strips on the Solids and Liquids class display to look like this:

7. **Circulate and note if any students are having trouble with the categories and definitions.** For students who don't read yet, read each definition aloud for them and have them check their classification of objects.

8. **End the class.** When time is up, have the students put their objects back in the bags. Collect the bags and the journals.

Students will continue recording on the same journal pages in Activity 2.

Reading through the journals will help you get an idea of their level of comprehension so far.

■ Going Further

1. **Drawing more solids and liquids.** For homework, tell the students to add a specific number of drawings of solids and liquids to their Solids and Liquids sheets, or to draw them on separate pages (so the journals can stay in class). Older students could make a list of solids and liquids instead.

2. **Bringing in more solids and liquids.** In Activity 2 you will be assembling a class display of solids and liquids. You may want to ask students to bring a solid and/or liquid to class to examine with their classmates and later add to the display. You could also make this a voluntary activity.

SOLIDS

You can tell something is a solid because:
• It holds its shape. (It doesn't make a puddle.)

Draw and label solids here:

Hard Solids

Soft Solids

LIQUIDS

You can tell something is a liquid because:

• It doesn't hold its shape. (It makes a puddle.)
• It takes the shape of its container.
• It stays flat on top, unless moving.

Draw and label liquids here:

ACTIVITY 2: COLLECTING SOLIDS AND LIQUIDS

Overview

Activity 2 begins with a review of the definitions of solids and liquids, and an opportunity for students to think of and share examples of solids and liquids. Next, pairs of students go to 10 learning stations around the room, collecting samples of different materials, often using interesting tools. They collect the samples in small bags, label the bags, and record in journals how they classified the samples. They then place each bag in the section of the Solids/Liquids class display where they think it belongs.

After assembling the display, the class discusses the placement of the items. Part of this discussion focuses on powders and granular substances, which many students confuse with liquids because they pour. The students are challenged to add true statements about solids and liquids to the class display. Each suggestion is discussed and evaluated by the group, using the evidence. Statements they agree on are added to the class display and to their journals.

Activity 2 may take one or two class sessions, depending on your students' age and experience, and your own preferences. If you present Activity 2 in one session, be sure to allow at least 15 minutes for the concluding discussion.

If you make it two sessions, stop the first session after the learning station activity, and use the second session for the concluding discussion and journal writing. The extra time in the first session helps make sure that all students have time to visit every station. The second session will allow ample time for students to "mine" their experiences, and for the discussion to reinforce key concepts and address common misconceptions.

Learning Objectives for Activity 2

Students will:

- reinforce their understanding of solids and liquids
- apply the definitions of solids and liquids
- debate whether collections of materials, such as powders or sand, are solids
- use simple tools to collect and measure substances
- base their explanations on evidence
- follow directions, work cooperatively in groups, discuss, and record data

Home Safety Reminder: You may want to tell the students that in science class they are observing and investigating only **safe** liquids and **safe** solid materials. Caution them not to do similar things at home unless they have full permission from their parents, who will make sure that they are investigating **safe** substances.

■ What You Need

For the class

- ❏ about 150 sealable, clear plastic snack-size bags
- ❏ the class display from the previous session
- ❏ several sentence strips
- ❏ 1 large wide-tip felt marker
- ❏ about 150 pushpins (or masking tape)

For the Learning Station Activity

- ❏ 4 cafeteria trays (to contain the mess at selected learning stations)
- ❏ 2 dishtubs
- ❏ 5 large plastic trash bags to cover tables or desks
- ❏ 12 plastic teaspoons
- ❏ students' **Solids** and **Liquids** sheets from the previous session
- ❏ pencils
- ❏ *(optional)* colored pencils or crayons
- ❏ 1 roll of masking tape or two sheets of address labels
- ❏ 6 black fine-tip permanent markers
- ❏ several sheets of aluminum foil
- ❏ *(optional)* 2–4 pairs of scissors
- ❏ 1 box of wooden toothpicks
- ❏ 2 tweezers
- ❏ 20 cotton balls
- ❏ 1 candle or chunk of wax
- ❏ 2 plastic serrated knives
- ❏ *(optional)* 2 craft sticks
- ❏ 4 rolling pins for crushing (or mortars and pestles, or wooden hammers)
- ❏ 1 box of chalk, any kind
- ❏ 2 collecting devices for liquid, such as medicine droppers, turkey basters, syringes without needles, sponges, spoons, or scoops
- ❏ 1 cup of water
- ❏ 5 drops of food coloring, any color
- ❏ about 2 cups of dry cereal that can be ground into a powder
- ❏ 1 hand-operated juicer
- ❏ about 3 oranges

■ Getting Ready

Before the Day of the Activity

Decide whether to recruit adult volunteers. If you have more than 20 students, and/or your students are not accustomed to working independently at centers or learning stations, we strongly recommend that you arrange for one or more parents to help during this session. Volunteers are also recommended to help with preparation and clean-up.

On the Day of the Activity

1. **Set out plastic bags.** Spread the empty plastic snack bags on a table or counter away from the stations and the Solids/Liquids display. There should be enough bags so each pair of students can pick up eight bags. Keep a few extra bags in reserve in case you need them.

2. **Have sentence strips, a large felt marker, and pushpins (or masking tape) near the class display.** Set the pushpins where they will not be knocked over.

3. **Write the steps of the procedure** in large letters on the board or chart paper posted where you will introduce the activity. Add a small sketch to illustrate each step, as suggested below.

It is best if groups get to post their bags on the display during the activity, and most students do fine with the pushpins. However, one first-grade teacher found that students were having trouble pinning bags up, which resulted in crowding at the display. She put out two buckets, one near the Solids side and one near the Liquids side of the display. During the activity, she had students put their bags in the appropriate bucket. Then, before the class discussion, she posted the bags on the display.

PROCEDURE

1. Get a bag.

2. Get a sample at a station. (about 1 teaspoonful)

3. Decide — solid or liquid?

4. Label the bag.

5. Record in your journal.

6. Put the bag on the wall.

4. Set up the ten learning stations around the room. Pairs of students go to stations and to the class Solids/Liquids display at their own pace. The activity is not a timed rotation.

a. Logistics and Room Arrangement

1. **Not much space is required at each station.** Depending on how many students you have, allow space for two to four students to work at a station at a time. Allow enough space between stations for students to move as freely as possible.

2. **Where to set up the stations.** If possible, put some stations on counters and at freestanding tables around which four students can stand at once. Try not to set up liquid stations in a rug area.

3. **Plastic trash bags for spills.** For the labeling station and the last four stations listed below, we recommend covering the table or desk with a large plastic trash bag. Cut the plastic bag so it is one big piece. Sprinkle a small amount of water on the table. Set the plastic bag on the wet table and spread it out by flattening it with your hands. The bag should adhere to the table.

b. Station by Station Set Up

1. **Drawing Station:** Set out all student journals with the Solids and Liquids sheets from the previous session and several pencils. *Optional:* Set out colored pencils or crayons.

2. **Labeling Station:** This is where the students will label the contents of their bags. Set up the labeling station near the drawing station. Cover the table with a plastic trash bag, because occasionally the bags leak when liquids are labeled. Set out masking tape or address labels and about six permanent markers.

3. **Aluminum Foil Station:** Set out several sheets of aluminum foil. *Optional:* Two pairs of scissors.

4. **Wooden Toothpicks Station:** Set out the wooden toothpicks and one or two plastic teaspoons for measuring the quantity.

5. **Cotton Balls Station:** Set out the cotton balls and two tweezers.

If space on tables and counters is limited, set up some or all of the stations at students' desks. Set up stations before class, and have the students go directly to a gathering area away from the stations for your introduction. Alternatively, you can have materials for the stations on trays off to one side, ready to distribute to desks and tables after your introduction.

For younger students who may have trouble writing the names of the samples, you could put a list of all the items at the station on a Word Wall. Or you might decide to have the students simply label the bags Solids or Liquids, or just S or L.

6. **Wax Station:** Set out the candle or chunk of wax on a cafeteria tray. Set out two plastic serrated knives (or craft sticks or pairs of scissors) and one or two plastic teaspoons.

7. **Chalk Station:** Break up the chalk into pieces about an inch long (allow one piece for each pair of students) and set them on a cafeteria tray. Set out two crushing devices (rolling pins, mortars and pestles, or wooden hammers). Set out one or two plastic teaspoons for measuring the quantity.

8. **Colored Water Station:** Pour about one cup of water into a dishtub. Add five drops of food coloring to the water. Set out two collecting devices (medicine droppers, turkey basters, syringes without needles, sponges, spoons, or scoops). Set out two plastic teaspoons.

9. **Cereal Station:** Set out the cereal on a cafeteria tray. Set out two crushing devices (mortars and pestles, rolling pins, or wooden hammers). Set out one or two plastic teaspoons.

10. **Orange Juice Station:** Set up the station next to a sink or set out a dishtub with water for washing hands. Set out one juicer. Cut each orange in half and set them out. Set out one or two plastic teaspoons.

For the following stations we recommend that you cover the table with a plastic bag. For the first three, add a cafeteria tray.

A Note on Setting Up Optional Stations

It is fine to use materials you have on hand to substitute for the stations suggested above or to create additional stations. **However, be sure your stations include a solid students can grind down to a powder (like chalk or dry cereal), some liquids, and both hard and soft solids.** It's also helpful to include wood, paper, and plastic, as they are such common materials. Some optional stations are:

• Vinegar Station: Set out a dishtub with some vinegar in it, droppers or turkey basters, and plastic teaspoons.

• Additional Colored Water Stations: Set out dishtubs with other colors of water, medicine droppers or turkey basters, and plastic teaspoons.

• Sponge Station: Cut a sponge in half and set it out with some scissors.

• Styrofoam Station: Set out a piece of Styrofoam, plastic serrated knives, and one or two plastic teaspoons.

• Plastic Toothpick Station: Set out some plastic toothpicks and one or two plastic teaspoons for measuring quantity.

• Fabric Station: Set out a piece of fabric and two pairs of scissors.

• Yarn Station: Set out some yarn and two pairs of scissors.

• Plastic Bag Station: Set out a few plastic bags and two pairs of scissors.

• Paper Station: Set out some paper towels, tissue paper, or cardboard and two pairs of scissors.

• Dry Pasta Station: Set out a rolling pin or wooden hammer, a piece of cloth to cover the pasta while pounding it, and one or two plastic teaspoons.

■ Reviewing Solids

1. **What is a solid?** Gather the students in a circle away from the learning stations. Ask, "How do scientists know if something is a solid?" [It holds its shape and doesn't make a puddle.] Emphasize that scientists don't make a decision to group an object as a solid or liquid just because it is their opinion.

2. **Review the definition of evidence.** Remind the students that scientists base their decisions on evidence. For example, "The rock is solid, because I set it down and it held its shape and didn't turn into a puddle."

3. **Ask the students to think of a solid.** Tell them everyone will get a turn to name one solid. Give them a minute to think about this, then ask them to raise their hands if they have one ready.

4. **Have them share.**

   a. Note which part of the circle has more hands raised, and begin calling on students in this area. This allows other students who might be confused to hear some ideas before their turn.

   b. Tell the students to put their hands down and then, one at a time, call on all students, working your way around the circle. Say it's fine if they want to share the same solid another student already shared. If a student is "stuck," come back to him later.

   c. As students share examples of solids, occasionally point out that each example *holds its shape and doesn't make a puddle.* This will help to reinforce the definition and the idea of backing up ideas with evidence.

   d. If a student brings up an example that is challenging, such as sand, shaving cream, or powder, ask if others can think of evidence of whether it is solid or liquid. Don't divulge yet whether it's a solid or not, but encourage them to keep thinking about it.

5. **Point to a solid.** After everyone has had a turn, tell them that on the count of three, every person will point to a solid somewhere in the room. [This should be easy!] Give them about 10 seconds to choose, then count to three.

6. **To reinforce the definition, quickly describe where a few students are pointing.** For example, say, "Ahmed is pointing at the chalkboard, which is holding its shape and not turning into a puddle. This is evidence the chalkboard is a solid."

If you gave the class a homework assignment to add solids and liquids to their journals, they could share one of these solids.

■ Reviewing Liquids

1. **Review the definition of liquids.** Ask, "How do we know if something is a liquid?" "What is our evidence?" [If we see that it doesn't hold its shape, makes a puddle, is flat on top except when moving, takes the shape of its container.]

2. **Tell each student to think of a liquid.** It doesn't need to be in the room. Remind them that anything they drink is a liquid (although there are many liquids they can't drink—like poison).

3. **As before, go around the circle,** calling first on several students who raised their hands, then on all the rest.

4. **Occasionally reinforce the definition of a liquid.** For example, "Leticia says spit is a liquid. Everyone use your tongue to feel the saliva in your mouth. Does it feel like it makes a puddle? Does it feel wet? If so, that is evidence that it is a liquid."

You may want to wait until just before they go to the learning stations to mention that they will be working with a partner. This way they can focus on the procedure instead of wondering who their partner will be!

■ Introducing the Solids and Liquids Learning Stations

1. **Tell the students they will soon go to learning stations** around the room, collect samples of solids and liquids, and add them to the class display.

2. **Model the procedure as they watch.** Explain that a *procedure* is all the steps used to do a task.

Step 1. Get a bag. Show where the bags are located. Emphasize that partners will take one bag at a time.

Step 2. Get a sample at one station. Tell the students less crowded stations are best, so they don't have to wait in line. There is no special order to the stations.

    a. Put a sample in a bag. Tell the students a *sample* is a small amount—in this case, about one teaspoonful. Emphasize that they will collect a small sample, not fill their bag.

    b. Use tools to gather samples. Show how to use the tools you think they might have difficulty with or any that might require safety instructions. (Don't model all the tools, because part of the challenge is figuring out how to use each one.)

    c. Carefully seal the bag.

Step 3. Decide if you think the object at the station is solid or liquid. Use your senses to find evidence of which it is. Talk it over with your partner.

Step 4. Label the bag. Go to the Labeling Station and demonstrate how to put masking tape or an address label on the bag, and write the name of the sample (for example, *chalk*) with a permanent marker. Don't write directly on the bag because it might poke a hole in the bag.

Again, for younger students, point out the Word Wall to help with spelling, or ask them to label more simply, for example, with S or L.

Step 5. Record in your journal. Point out their journals at the Drawing Station. Emphasize the importance of drawing or writing the name of the sample in the appropriate place on their Solids or Liquids sheet.

Step 6. Put the bag on the wall. Show how to stick a pushpin into the bag above the seam and stick it to the display. Make sure the students understand they need to classify by posting a bag on the Solids or Liquids side. If someone else has already put a bag with the same kind of sample on the class display, it is fine to add theirs as well.

Tell the students that once they have put their bag on the display, they can get another bag and go to a different station. Remind them that if they forget what to do, they can refer to the procedure you have posted.

4. **Review the procedure by modeling the wrong way.** Tell the students you're going to review the steps again, this time by doing everything the wrong way. It will be their job to tell you what you are doing wrong and what would be the right way. (See example.)

A Sample Wrong Way Demonstration

Step 1. Be rude at the bag table.

- Tell your partner to go away because you don't feel like working with him.

- Boss your partner around, saying, "OK partner, I'll work with you. You do everything I tell you, and we'll get along just fine."

- Take all the bags from the bag table.

Step 2. Be careless and rude collecting a sample.

- Run with scissors.

- Squirt a liquid in the air.

- Take a piece of equipment away from another student who is using it.

- Crush all the chalk on the tray. Spill the chalk off the tray and onto the floor.

- Put a whole object (such as a sheet of foil) in your bag.

- Don't seal the bag well, and hold it upside down.

Step 3. Say you think something is liquid without even looking at it closely.

Step 4. Write directly on the bag rather than on a label (which often pokes holes in the bag).

Step 5. Don't use your journal correctly.

- Forget to draw in your journal before going to the display.

- Draw a liquid on the Solids section of your sheet or vice versa.

Step 6. Put the bag on the display without a label.

---

4. **Set a signal for getting the students' attention.** Tell the class you will flick the lights (or whatever signal you use) to get their attention if necessary. Mention that you will give a five-minute warning before the activity ends, so they can finish the station they are on and help tidy up the stations.

5. **Assign pairs to their first stations.** Emphasize that they should stay with their partners and work together. Remind them that partners will share a bag for each station. Hand out the first bag to each pair and assign them to their first station, to make sure the class is evenly distributed.

■ Conducting the Learning Station Activity

1. **Circulate as the class is working, helping out as needed.** Notice if any students are having trouble with the categories and definitions.

2. **Be sure that both partners record in their journals before going on to a new station.** If they rush, encourage them to slow down and do a good job. Say that it is not necessary to go to all the learning stations (although they will probably want to).

3. **Five-minute warning.** Give a signal to let the students know they have five minutes left. Ask a few pairs to tidy stations as the rest of the class returns to the discussion area.

■ Discussing Collections on Display

1. **Gather the students where they can see the class display.** Compliment them on the work they did.

2. **Discuss classifications.** Ask, "Are there are any objects on the display that you were not quite sure how to sort?" Encourage discussion, and let the class decide whether or not to re-categorize particular samples. When possible, encourage students to explain their thinking by asking, "What did you see that makes you say that? What is your evidence?"

3. **Discuss chalk dust and crushed cereal.** Some students may classify powders as liquids because they "pour." If no one brings up this idea, bring it up yourself and allow some time for discussion. You can hold off until the next session, "Challenging Substances," to make a final decision with the class.

One teacher said, "I was so impressed! They worked in pairs wonderfully—took turns, discussed. This was wonderful!"

If you did decide to break Activity 2 into two sessions, this is a good point to end Session 1. It is important to allow at least 15 minutes for a concluding discussion, to make sure the students have time to process their ideas and experiences.

See page 40, "A Sample of a Class Discussion in Activity 2" for an example of how such a discussion might go and ways that teacher-student interaction and questioning can deepen understanding.

The idea that a powder is a liquid is a common misconception.

Pouring is not really a definition of a liquid, because liquids, gases, and tiny solids can all be poured.

For your own information, see the Background for the Teacher section (page 88) for more complete scientific definitions of solids and liquids, and how scientists classify some challenging substances.

■ Adding to the Definitions of Solids and Liquids

1. **Ask for more definitions for the display.** Tell the students you'd like their help in adding to the definitions of solids and liquids.

2. **Ask, "Is there anything else that is true of all solids or all liquids?"** As an example, tell them they can't say, "All solids are red," because only some solids are red. They can say, "All solids hold their shape," because they all do.

3. **If needed, give clues by asking questions,** such as, "Can you smash solids?" "Can you smash liquids?" "Tell me about splashing." "Do solids feel wet?"

4. **Post new definitions.** Discuss each suggestion they make, and if everyone can agree, write it on a sentence strip and post it on the display.

5. **As possible, illustrate the sentence strips with simple drawings,** especially if you have pre-readers in your class.

A common tricky one is the chalk dust, because, as a powder, it pours, somewhat like a liquid. It's common for students to mistakenly classify powders and other tiny solids that pour as liquids. That's why the chalk or cereal stations are important to include, because they allow students to take a larger piece of chalk or cereal that they recognize as a solid, then break it down into tiny pieces (dust). Hopefully they will then recognize that the dust is the same solid substance, just broken down into smaller solid pieces. Encourage students to discuss and debate its placement without necessarily reaching agreement. You can hold off until the next session, "Challenging Substances," to make a final decision.

Some examples of definitions your students might offer during this discussion or later in the unit:

Solids

Can be any color and any shape.

Stay the same size.

Have mass (or weight).

Take up space.

Can be powder.

You can't put your finger through them without breaking them.

Liquids

Can't be smashed or cut.

Can be any color.

Can be the shape of any container.

Have mass (or weight).

Take up space.

Are not powder.

■ Recording in Journals

1. Have the students add the new definitions on the display to their journals at the top of their Solids and Liquids pages.

2. *Optional:* **Have them do a "quick write" in their journals** about their evidence for whether they think chalk (or cereal or whatever similar substance you used for crushing) is a solid or a liquid. Make sure they understand that they don't have to be right—writing this is just a way of recording what their current thoughts are and why they think what they do.

Your Solids and Liquids class display should now look something like the picture below.

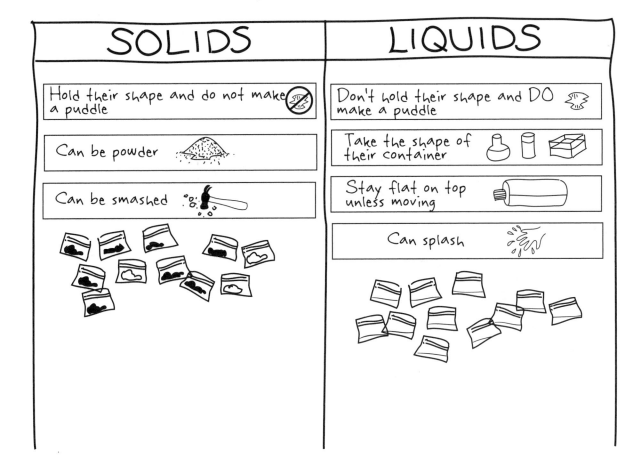

Sample of a Class Discussion in Activity 2

Student A: You can put your finger through liquids.

Teacher: Shawna has suggested adding, "You can put your finger through it" to our liquid definition. Raise your hand if you've ever stuck your finger through a liquid. Hmm, it looks like a lot of people have noticed the same thing. Can anybody think of any liquids you can't stick your finger through?

Student B: You can't stick your finger through lava, cause it'll burn.

Student C: You can't stick your finger through acid.

Teacher: OK, it seems like there are some liquids you can't stick your finger through.

Student D: Maybe we can say that you can stick a solid through it.

Teacher: What do you all think of the statement "You can stick a solid through it"?

Student B: Cotton candy is solid, but you can't stick it through a liquid cause it'll dissolve.

Student A: A finger is a solid and you can't stick your finger through acid.

Student C: Maybe we can say, "You can stick a metal pole through it."

Student B: But even a metal pole will melt in lava.

Teacher: Sounds like we don't yet have a statement we can add. Keep thinking about this over the next few days. Maybe someone can think of a way to say it so we can all agree. Does anyone have another suggestion?

Student: D: If you step in liquid, it splashes.

Teacher: Could we shorten that and say, "Liquids splash"?

Student D: Dust is solid, and if you step in it, it splashes.

Teacher: Actually, we use the word splashing for liquids. When dust goes up it's not splashing, but it's kind of like splashing. What might we call that?

Student B: The dust jumps up?

Teacher: That might work. Let's get back to "Liquids splash." Can anyone think of liquids that don't splash? Raise your hand if you don't think we should put this definition under liquids.

(No response.)

Teacher: So let's write "Liquids splash" on a sentence strip and add it to the liquids definitions.

Teacher: Can anyone think of another thing we can say about either solids or liquids?

■ Going Further

1. **Highly recommended: Solid and Liquid Scavenger Hunt.**
   Either as homework or as additional class work, have students search
   for various solids and liquids (for example, a solid that is soft, a liquid
   you can't see through but that light will go through). They then name
   and/or draw an example of each. See page 42 for the optional **Solid
   and Liquid Scavenger Hunt** sheet.

2. **Journal Entries.** For homework or class work, assign each student to
   add at least five more solids and liquids to her journal.

3. **Adding to the Display from Home.** With parents' permission, have
   students bring in solids and liquids from home to add to the display.
   You can also add other items from around the classroom.

Solid and Liquid Scavenger Hunt

You can tell something is a solid because:

— It holds its shape (it doesn't make a puddle).

Find these solids. Name or draw them on your sheet:

• a solid that is hard _____

• a solid that is soft _____

• a solid that has a shape you can change _____

• something that's hard to tell if it's solid or liquid _____

• a lightweight solid _____

• a heavy solid _____

• a solid that turns into a liquid _____

You can tell something is a liquid because:

— It doesn't hold its shape (it makes a puddle).

Find these liquids. Name or draw them on your sheet:

• a thick liquid _____

• a runny (thin) liquid _____

• a liquid that you can't see through, but light will go through _____

• a liquid light won't go through _____

SOLIDS

You can tell something is a solid because:

• it holds its shape (it doesn't make a puddle).

Draw and label solids here:

Hard Solids   **Soft Solids**

chalR dust

tooth picks

pebles

beed   paper clip

corn puffs

cloth   coton ball

fur

cotton

Sample of Student Work

Overview

At learning stations pairs of students explore four new substances—shaving cream, toothpaste, sand, and "Glook." Each of these substances is a challenge to categorize as a solid or liquid. The students apply their definitions of solids and liquids, and record their thinking on their student sheets.

After the investigations, the class gathers for discussion. The students are introduced to hand signals for solids and liquids that allow everyone to share ideas and give the teacher an idea of what students are thinking.

Activity 3 builds thinking and discussion skills and models the processes scientists use to grapple with things that don't fit neatly into categories. Another goal of this activity is to deepen students' understanding of the definitions of solids and liquids, especially the idea that granular substances are solids. Even after their deliberations about chalk and cereal in Activity 2, many students have difficulty giving up the idea that small-grained substances are liquids because they can be poured. Guided discussion and journal writing help students overcome this common misconception.

Glook is like silly putty, but softer and not as rubbery. See the Glook Station under "Getting Ready" (page 47 and 48) for directions on mixing Glook.

Learning Objectives for Activity 3

Students will:
- reinforce their understanding of solids and liquids
- understand that materials such as powders and sand are solids
- apply definitions of solids and liquids
- base their explanations on evidence
- use simple tools to collect and measure substances
- follow directions, work cooperatively in groups, and record data

---

■ What You Need

For the class
- ❏ Solids and Liquids display from previous sessions
- ❏ rock and a few other solid and liquid items from previous sessions
- ❏ a small amount of baking soda
- ❏ 1 clear plastic cup

For each student
- ❏ journal from previous sessions
- ❏ one copy of the **Solid or Liquid?** student sheet (page 54)
- ❏ for pre-writers: sticky notes instead of a student sheet
- ❏ pencils

For the toothpaste, shaving cream, and sand stations (two of each)
- ❏ 4 cafeteria trays
- ❏ 4 trash containers
- ❏ about a cup of shaving cream
- ❏ about $\frac{1}{2}$ tube of toothpaste
- ❏ toothpicks
- ❏ about a cup of sand
- ❏ 6 wide-mouth plastic cups
- ❏ 8 hand magnifying lenses OR handheld microscopes★
- ❏ 4 dishtubs for washing hands
- ❏ paper towels
- ❏ water
- ❏ 2 dustpans and brooms for cleanup
- ❏ *(optional)* dissecting microscopes

For two Glook stations
- ❏ a 1-teaspoon measuring spoon
- ❏ a 1-tablespoon measuring spoon
- ❏ 2 sealable plastic bags or other airtight containers
- ❏ 8 tablespoons of white glue
- ❏ 8 tablespoons of water
- ❏ 2 cups
- ❏ 1 stirrer
- ❏ 4 teaspoons of Borax powder *(Sodium Tetraborate, used in laundering and as a household cleaner)*
- ❏ 1 cup of warm water

Any slimy substance that is safe and challenging to classify as a solid or liquid, such as hair gel, can be substituted for the toothpaste or shaving cream.

***About handheld microscopes:** These are for the sand station. Flashlight-style 30x handheld microscopes are optional, but work very well. They are usually powered with two AA batteries. They sell for about $10 to $20 in electronics and "nature" stores, in many catalogs, and online.*

*Glook is a little messy to make, but not messy once it's made, so trays, trash bags, and hand washing are **not** needed for students.*

■ Getting Ready

Before the Day of the Activity

1. **Decide if you will have the students record their ideas on the Solid or Liquid?** student sheet (page 54). If so, make a copy for each student.

2. **For pre-writers,** you may not want to use the student sheet. Instead, write the names of the substances on sticky notes and place them at each station. Students can then place a sticky note in either the Solids or Liquids section of the class display.

3. **Have available** near where you will hold class discussions:
   — a few solid and liquid items from previous sessions
   — a small amount of baking soda in a clear plastic cup
   — **Solid or Liquid?** student sheets
   — journals from previous sessions

On the Day of the Activity

Set Up the Learning Stations

1. **Shaving Cream Stations:** Squirt some shaving cream into a plastic cup. Set it on a cafeteria tray. If the stations are not near a sink, set out a dishtub partially filled with water and some paper towels. Place a trash container nearby.

2. **Toothpaste Stations:** Squirt some toothpaste into a container. Set it on a cafeteria tray to contain the mess. If the stations are not near a sink, set out a dishtub partially filled with water and some paper towels. Place a trash container nearby.

3. **Sand Stations:** (Students examine sand directly on the table.) Put some sand in a plastic cup and set it out. Set out the magnifiers or handheld microscopes. Place a broom and dustpan nearby.

4. **Glook Stations:** For one batch of Glook: Put 4 tablespoons of white glue and 4 tablespoons of water in a cup and stir. In a different cup, put 2 teaspoons of Borax powder in a cup of warm water, and stir until it dissolves. Pour the glue solution into the cup with the Borax solution and stir. A glob of slime should start to form. Poke any bubbles with your stirrer. After stirring it for a while, take the glob out and use your fingers to massage and mix the ingredients together more. Do this until the Glook no longer feels wet. It should be malleable like silly putty, but softer and not as rubbery.

There are four substances to investigate. To provide plenty of room for student pairs to work, we recommend you set up eight learning stations—two for each substance. If each station accommodates four students at a time, there will be space for 32 students. For smaller classes, adjust accordingly.

Repeat the process for the second batch of Glook. Store the Glook in a sealable bag, to prevent it from drying out. Just before class, remove the Glook from the bag and set it out directly on the surface of the Glook station.

■ Introducing Solid and Liquid Hand Signals

1. Introduce the class to the following hand signals:
closed fist = solid
slowly waving hand = liquid

Some teachers tell their students to hold their signals close to their chests so they will be less likely to copy other students.

Tell the students they should keep making their signal long enough for you to see what each of them thinks. Ask them not to say the word while making the signal. (You may need to remind them about this.)

2. Hold up a few liquids and solids and ask for hand signals.

3. Hold up one of the containers with liquid in it. Ask the class to show their hand signal for whether *everything* you're holding in your hand *(hint, hint)* is solid or liquid. If anyone holds up the solid signal, or both the liquid and solid signals, ask her why. [Because the stuff inside the container is liquid—it doesn't hold its shape, and takes the shape of the container—but the container itself does hold its shape, which makes it a solid.]

■ Introducing Challenging Substances

1. Draw attention to the class display. Tell the class they're doing very well acting as scientists, sorting and classifying things in the Universe into solids and liquids.

Let the students grapple with the definitions as they try to classify these challenging substances. In the concluding class discussion, they will learn that sand is a solid, but the other three substances fit neither the solid nor liquid categories.

2. Are shaving cream, toothpaste, Glook, and sand solids or liquids? Let the students know that the four things they will work with today may be hard to classify as solids or liquids. Say that, like scientists, they should talk with their partners, think about the definitions, get all the evidence they can with their senses, and do their best.

3. Tell the students they will be going to stations with a partner. They can go to the stations in any order. **This time they will not collect samples.** They will try to figure out if each substance is a solid or a liquid and record their decision on a sheet.

4. Only four stations. Make sure the students understand that there are eight stations set up to make room for everyone, but they will go to only four of them because there are two of each station.

■ Introducing the Procedure

1. Pass out a Solid or Liquid? sheet to each student. As you go over the steps of the procedure, point out where on the sheet they will record. Say they should bring their paper and pencil to each station.

2. Demonstrate the steps for each station.

   a. Go to a station with a partner. First, look for evidence about the sample using the senses of touch, sight, and smell. **No tasting!**

   b. Check the list on the class display (or in the journal). Read the **entire** list to decide if the toothpaste is solid or liquid. Students should discuss their thinking with their partners.

   c. Circle *Solid, Liquid,* or *Not Sure* on the student sheet, and write why.

   • If the sample matches everything on the liquids list and nothing on the solids list, they will circle *Liquid* on their sheet, and then write why they think it is a liquid.

   • If the evidence matches all the solid definitions and no liquid ones, they'll circle *Solid* and explain why.

   • If a sample matches some solid and some liquid definitions, they should circle *Not Sure* on their data sheet and explain why.

3. Set guidelines for containing the mess. As the students watch, go from station to station and demonstrate some rules designed to keep the mess under control:

   a. **Keep the shaving cream and toothpaste on the trays.** It is okay to put the sand and Glook on the table, but try to keep them off the floor.

   b. **Use the broom and dustpan to clean up spills at the sand stations.** It is okay to put some sand on the table. Demonstrate how to use the magnifiers (or handheld microscopes).

   c. **At the toothpaste stations, take one toothpick** and pick up a small amount of toothpaste to examine. Put the toothpick in the trash container when finished.

Beginning writers can just circle Solid, Liquid, or Not Sure on their sheets. Later, during the class discussion, they can explain their reasoning verbally.

Option for pre-writers: Explain that they will put a sticky note on the class display instead of writing on the sheet.

Discussing Glook, shaving cream, and toothpaste helps students demonstrate and further strengthen their understanding of the definitions of solids and liquids. However, correctly classifying those three substances is not the main goal of this activity.

Understanding that sand is a solid is a more fundamental goal. Therefore, if you do not have enough time for a thorough discussion of all four substances, make sure you allow enough time to discuss sand (see page 51).

Colloids can also be a mixture of solids and gases, as in the case of shaving cream. But it's best not to discuss gases with students until after Activity 5, when they will have explored gases. In this session, you may want to mention that colloids are a mixture of two or more things so they can't be categorized as either a solid or a liquid.

In contrast, the next substance the class discusses, sand, is not a combination of more than one thing, so it can be categorized. While it's fine to mention colloids to students, understanding colloids is beyond the scope of this unit. Please see Background for the Teacher for your own interest.

d. **At the shaving cream stations, use only your "pointer finger"** to pick up a small amount of shaving cream.

e. **Wash your hands after exploring the toothpaste or shaving cream.** Point out dishpans or sink.

■ Conducting the Learning Station Activity

1. **Pass out a Solid or Liquid? sheet to each student.** Have them write their names at the top.

2. **Assign pairs to their first station, and let them begin.**

3. **Circulate and be sure the students are applying the class definitions** as they classify solids and liquids. Remind them to record their reasoning before moving on to the next station.

4. **Five-minute warning.** When most students have been to most stations, have the class "freeze," and tell them they have five more minutes. Explain that it's OK if they don't get to all four substances, but they should try to get to a sand station if they haven't yet.

■ Discussing Glook, Toothpaste, and Shaving Cream

1. **Briefly discuss Glook.** Gather the students, with their sheets, away from the stations. Ask who circled *Solid* and why. Do the same for *Liquid* and *Not Sure.* Encourage discussion of alternate opinions.

2. **Discuss toothpaste and shaving cream.** Validate the students' thinking as they discuss how the definitions do or do not apply to these substances.

3. **Neither solid nor liquid.** Tell the students that they did a good job of thinking and using the definitions. Summarize by saying that other scientists agree—these three are tricky to sort into solids or liquids!

4. **Reveal that scientists classify these three as *colloids*.** Tell the students that toothpaste and Glook are tiny pieces of solids mixed into liquid. Scientists call a substance like this a colloid. Mention that shaving cream is a colloid too.

5. **Discussion is valuable.** Confirm that toothpaste, shaving cream, and Glook are not really solids or liquids, but you wanted the class to have a chance to discuss them and think about the definitions—just like scientists do.

■ Discussing Sand

1. **Ask for hand signals about sand.** Hold up a sand sample, and ask if it is solid or liquid. Mention that, unlike the Glook, toothpaste, and shaving cream, sand does fit into one of the categories.

2. **Ask students to share evidence they found that sand is a solid or liquid.**

3. **Guide them to an understanding that each grain is a solid.** If students say that sand appears to flow or pour like a liquid, focus on individual pieces of sand, and how each one exhibits properties of solids. Ask:

   a. "What is sand is made out of?" [Pieces of rocks or shells.]

   b. "What did the sand look like through the magnifiers?"

   c. "Did each piece of sand hold its shape?"

4. **Remind them of the chalk in the previous session.** Ask if a big piece of chalk is a solid. Ask, "When the chalk was broken up into medium-sized pieces, was it still chalk?" "Was it still a solid?" "Was it still a solid when it was broken up into small pieces?" "Tiny pieces?" "Powder?"

5. **Piles of tiny solids are still solid.** Emphasize that, even with a change in size, a solid is still solid. Even though they are tiny, each piece still holds its shape.

■ Explaining to Someone Why Powders Are Solids

1. **Hold up the cup of baking soda.** Say it is like sand, except the pieces are even smaller, and you would need a more powerful magnifier to see them and get evidence that they are solids.

2. **Pretend you are a person who thinks baking soda is a liquid.** Ask the students to explain to you what may be incorrect about each of the following statements:

   a. "It seems like baking soda takes the shape of its container." [Each tiny piece actually holds its shape.]

Remind the students to make a fist for solid or a slowly waving hand for liquid.

Some teachers like to ask students to imagine what grains of sand would be like to a tiny creature, like an ant. Ask the students to imagine what sand would seem like if sprinkled on top of them at that size. [It would be like boulders!] Point out that if an ant were to touch a grain of sand at that size, it would be like a person touching a hard boulder—because the grains of sand are solid and hold their shape.

It may be helpful to point out the difference between considering a whole container or pile of something and its individual pieces. For example, sand, when considered as a pile of sand, does act in some ways like a liquid, as opposed to when considering each individual grain, which makes it a solid. When a scientist categorizes sand, she has to think about each individual grain, similar to the way we'd think about each rock in a pile rather than the pile all at once. If a student is thinking only about sand as a pile or how all the sand in a container behaves, it can be helpful to point out this difference. The idea is not to say that the student is wrong, but to recognize the logic in his idea, and to help all the students understand that when scientists study substances like sand, they focus on the individual grain.

b. "It seems like you can stick your finger through baking soda." [You're actually sticking your finger *between* the tiny solids, not through them.]

c. "It seems like it makes a puddle." [Each tiny piece holds its shape and does not make a puddle.]

d. "It pours." [Liquids are not the only things that can be poured.]

■ Recording in Journals

Have the students add the **Solid or Liquid?** sheet to their journals. Give them time to write in their journals. Have them pretend they are writing a letter to someone who doesn't believe sand is a solid. How could they convince that person that it is a solid? What is the evidence?

■ Going Further

1. ***Two Bad Ants.*** Read the book *Two Bad Ants* by Chris Van Allsburg (Houghton Mifflin, Boston, 1988). In the story, two curious ants set off in search of beautiful sparkling crystals (sugar). Illustrations are drawn from an ant's perspective, showing individual sugar crystals, allowing students to see the solid properties of grains of sugar.

2. **Writing a "small" story.** Have your students write and illustrate their own story of a tiny creature's adventures in grains of a powder or sand. They could even write stories starring shrunken versions of themselves.

3. **Making Glook.** Use the recipe on the Meet Glook! sheet (page 55) and have students work in pairs to make individual batches of Glook they can take home in sealable sandwich bags. Be sure to cover the points—especially the safety points—on the sheet with your students before sending it home, and make sure the sheet is stapled to each sample for parents to read.

4. **Making Oobleck.** Using one of the two recipes on the next page, challenge your students to explore Oobleck and decide if it's a solid or a liquid. It is an intriguing substance that in some instances doesn't hold its shape, like a liquid, and in others is hard like a solid. *Please note that because of the many different grades of cornstarch, it is impossible to provide an exact recipe.* At the right consistency, Oobleck should flow when you tip the bowl, but feel like a solid when you hit it or rub your finger across the surface. If it is too thick to flow, add a little water. If it is too soupy, add a little more cornstarch.

Oobleck recipe for individuals: Put about $\frac{1}{4}$ cup of water into a cup. Mix in one drop of food coloring. Add about 9 teaspoons of cornstarch. Use your fingers to mix them together until smooth.

Oobleck recipe for the whole class: Mix 6 $\frac{3}{4}$ cups of water with about 15 drops of food coloring. Slowly sprinkle in four boxes of cornstarch. Mix by slipping your fingers under the mixture and "lifting" from the bottom of the bowl to the top by slipping your fingers under the mixture, until an even consistency is reached.

The GEMS guide Oobleck: What Do Scientists Do? *(Grades 4–8), uses this strange substance to explore attributes of matter and convey some essentials of scientific investigation.*

Solid or Liquid?

Explain your reasons.

Shaving Cream: Solid   Liquid   Not Sure

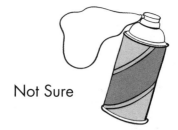

_____

_____

_____

Toothpaste: Solid   Liquid   **Not Sure**

_____

_____

_____

Glook: Solid   Liquid   Not Sure

_____

_____

_____

Sand: Solid   Liquid   Not Sure

_____

_____

_____

Meet Glook!

Glook is a substance the students observed during their study of solids and liquids. They were challenged to decide whether it's a liquid or a solid. Glook is made from Borax, white glue, and water. The result is safe to handle. However, your children know never to put chemicals that are not food in their mouths.

Your child has agreed to the following rules for Glook:

1. Keep Glook away from younger children and animals. (They could choke on it.)

2. Do not throw Glook. (An animal could find it and eat it.)

3. Follow the directions of your parents.

4. Please don't put Glook in your bed. Yuck!!

Recipe for Glook

1. Put one tablespoon of white glue and one tablespoon of water in a cup and stir.

2. In a different cup, put one teaspoon of Borax powder and a quarter cup of warm water, and stir until it dissolves.

3. Pour the glue solution into the cup with the Borax solution and stir. A glob of slime should start to form. Poke any bubbles with your stirrer.

4. After stirring it for a while, take the glob out and use your fingers to massage and mix the ingredients together more. Do this until the Glook no longer feels wet. It should be like silly putty, but softer and not as rubbery.

5. Explore it!

6. When not playing with it, keep the Glook in a sealable bag, to prevent it from drying out.

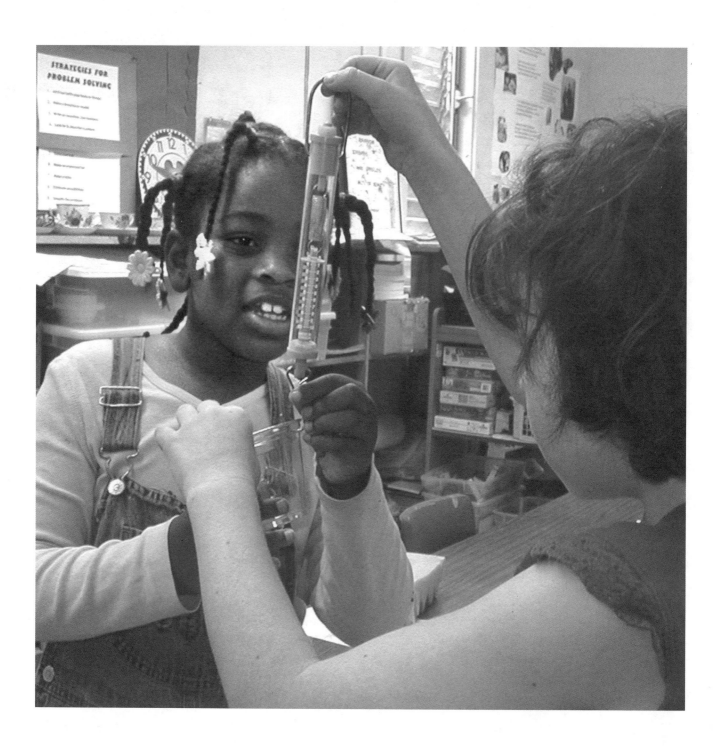

Overview

The students are introduced to the term *matter* and a simplified definition of it: that matter is anything in the Universe that has mass and takes up space. This definition and two new signs— Matter and Not Matter— are added to the class display.

With a partner, students explore a variety of objects, once again gathering evidence. They make predictions, use a spring scale to determine whether each item has mass, and use their fingers to test if it takes up space. In a class discussion, they share their results. They realize that all solids and liquids are matter. Through a demonstration with a balloon filled with carbon dioxide gas, the students are introduced to the idea (to be greatly expanded upon in Activity 5) that gases are matter too.

Finally, the students are challenged to think of things that are not matter. Those things that are agreed upon are added to the class display. The students also test whether people are matter, and conclude that yes, they are.

Learning Objectives for Activity 4

Students will:

- learn that matter takes up space and has mass
- apply the definition of matter by testing a variety of objects
- use spring scales to determine if objects have mass
- learn that all solids and liquids are matter
- support arguments with evidence
- follow directions, work cooperatively in groups, and record data

---

*Why use the term **mass** with young students? At Grades 1–3, it is too early for most students to understand the difference between weight and mass. However, there is a difference, and later in the students' education, it will be an important distinction to understand.*

Mass is a more universal, useful, and scientific word than weight to use in defining matter. Weight depends on where you measure it. On the Moon, an object would weigh less than it does on Earth, because the Moon has less gravity than Earth. But an object would have the same mass on Earth as on the Moon. The object would be made of the same amount of stuff, no matter how much it weighed. (Please see the Background for the Teacher section for more on mass and weight.)

Unless you have some very sophisticated third graders, don't try to explain this distinction. Just use the word mass yourself during the activity, and don't worry too much if your students continue to use the terms mass and weight interchangeably. (This is understandable because on Earth, we usually measure an object's mass by weighing it.)

■ What You Need

For the class

❑ Solids and Liquids class display from previous sessions
❑ 1 calibrated spring scale (up to 100 grams)
❑ about 12 sentence strips
❑ 1 black wide-tip marker
❑ 1 indelible marking pen (to mark spring scales)
❑ 1 lightweight object, such as a pencil
❑ 1 heavy object, such as a ring of keys
❑ 1 very heavy object, such as a large rock
❑ 2 identical balloons
❑ materials to make carbon dioxide gas (see "Getting Ready," Step 6, page 60)
   — 1 tablespoon vinegar
   — 1 teaspoon baking soda
   — a 1-tablespoon measuring spoon
   — 1 vial or pill bottle
   — 1 funnel

For each pair of students

❑ 1 spring scale (see A Note About Spring Scales, page 59)
❑ 1 fold-top plastic sandwich bag
❑ 1 paper clip or binder clip

For each student

❑ journal from previous sessions
❑ 1 copy of the **What's the Matter?** student sheet (page 67)
❑ 1 pencil

For the matter testing stations (see note)

❑ 1 piece of wood
❑ 1 large, lightweight object, such as a piece of Styrofoam
❑ 1 small, heavy object, such as a fishing weight
❑ 1 piece of plastic
❑ water (in a tied-off balloon)
❑ 1 sheet of paper
❑ 1 small rock
❑ 1 piece of glass (not sharp!)

It is fine to substitute other items for those listed here. Try to include a variety of materials. Make sure each item is heavy enough to register on the scales that the students will be using, but not too heavy. If you have a large class, provide two samples of some of the items, or add things like bottle caps, erasers, or anything you have handy.

A Note About Spring Scales

In this activity, pairs of students use a spring scale to weigh objects to determine if they have mass or not. Since the point of the activity is simply to determine whether or not an object *has mass*—**NOT** *how much mass* it has—**the scales don't need to be calibrated.**

In field tests of this activity, students used inexpensive, uncalibrated spring scales from Carolina Biological Supply Company. These scales have a long red spring encased in a transparent tube. Most teachers found them to be sensitive enough for this activity, and the low cost is an advantage, since you need one for each pair of students. If you purchase the GEMS Kit for this unit from Carolina, 16 of these scales are supplied along with one calibrated spring scale for your class demonstration.

Some teachers prefer to borrow or purchase traditional, calibrated 100-gram spring scales for the students as well. With these scales, you have the option to have older students record the mass of each object in grams, or the weight of each object in Newtons if the scale shows only weight. For more on mass and weight please see Background for the Teacher.

■ Getting Ready

1. Attach a sandwich bag to a spring scale for each pair of students.

  a. Using a paper clip or binder clip, attach the plastic sandwich bag to the hook at the bottom of the scale. The bag will be used to hold the items that students test.

  b. If you are using the uncalibrated red-spring scales, use the indelible marking pen to make a mark on the tube as a reference point for the students. With the sandwich bag attached, hold up the scale and draw a line on the tube near the metal at the bottom of the spring. The mark will help the students gauge whether the spring stretches out even when very light objects are being tested.

2. Prepare the calibrated spring scale for your demonstration. Toward the end of the session, you will weigh two balloons on the calibrated spring scale. One balloon will be filled and one empty. Decide how you will attach the balloons to the scale. A paper clip through the edge of each balloon may work well.

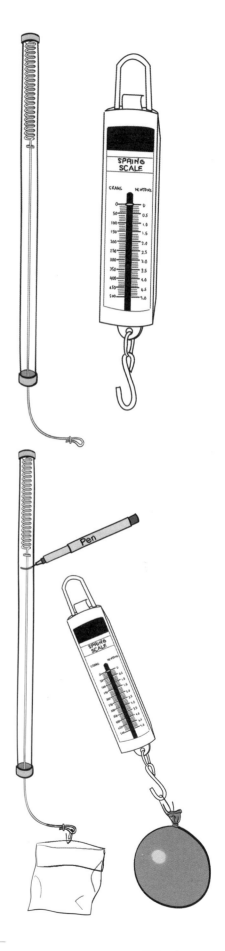

3. Make two large signs for the class display:

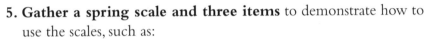

Matter

Not
Matter

Don't post the signs, *but have them near the Solids/Liquids display, along with a few more blank sentence strip signs and a wide-tip felt marker.*

4. Make six more signs on sentence strips:
- have mass
- air
- do not have mass
- take up space
- people
- do not take up space

5. Gather a spring scale and three items to demonstrate how to use the scales, such as:
- a pencil
- a ring of keys
- a large rock

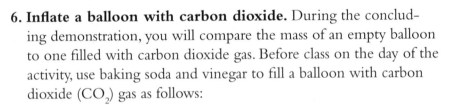

vial

6. Inflate a balloon with carbon dioxide. During the concluding demonstration, you will compare the mass of an empty balloon to one filled with carbon dioxide gas. Before class on the day of the activity, use baking soda and vinegar to fill a balloon with carbon dioxide (CO_2) gas as follows:

a. Stretch the balloon once or twice by blowing it up and then letting the air out.

b. Put a funnel into the mouth of the balloon.

c. Put a teaspoon of baking soda into the balloon.

d. Take the funnel out.

e. Put about a tablespoon of vinegar into a vial (or pill bottle).

f. Stretch the opening of the balloon over the mouth of the vial **without spilling the baking soda into the vial.**

g. Lift the end of the balloon so the baking soda falls into the vial. The balloon should inflate with carbon dioxide gas.

h. Take the balloon off the vial and quickly tie the balloon.

i. Have the carbon dioxide-filled balloon handy for your concluding demonstration, with an identical empty balloon and a spring scale.

7. **Set out objects for the matter testing stations:** Choose desks or tables around the room where student pairs can go to test objects made of various materials on their spring scales. Each station should have one or two items for students to investigate:

- wood
- large piece of Styrofoam, or other large, lightweight object
- small piece of metal or other small, heavy object
- small piece of plastic
- water (in a tied-off balloon)
- paper
- a small rock
- a piece of glass (not sharp!)

8. **Copy student sheets.** Depending on the objects you've gathered, you may want to add or delete items listed in the left column of the **What's the Matter?** student sheet (page 67). Make one copy of the sheet for each student.

9. *Optional:* **Make a sign for each station.** Make a sign with the name of each object to be tested. The names should match those on the **What's the Matter?** sheet.

■ Introducing Matter

Testing If Objects Take Up Space

1. **Introduce matter.** Tell the class that some things in the Universe belong in a group called *matter.* Show them the large Matter sign and place it so they can see it, but don't add it to your Solids/Liquids display yet.

2. **Say for a thing to be matter, it must take up space.** Put the "take up space" sign near the Matter sign.

3. **Ask, "Does a pencil take up space?"** Show how you are able to put your fingers together when the pencil is not between them.

4. **Pinch a pencil between your fingers.** Show that when you try to put your fingers together with the pencil between them, the pencil blocks them. This is evidence that the pencil is taking up space.

5. **Say that they will use this simple "finger test" on objects** to see if they are matter by checking to find out if they take up space.

Testing If Objects Have Mass

1. **Tell the students that to be matter, something has to have** *mass.* Mass means it is made out of "stuff."

2. **Put the "have mass" sign near the Matter sign.**

3. **Ask the class how they could test a pencil to see if it has mass.** Accept all ideas.

4. **Introduce gravity.** Hand someone the pencil to feel how heavy it is. Explain briefly that gravity is a pull between any two objects, in this case the Earth and the pencil. Gravity is pulling the pencil toward the ground. If the pencil did not have any mass (if it were not made of "stuff") gravity would not pull on it, and it wouldn't feel heavy.

5. **Test for mass using a scale.** Because of Earth's gravity, we can test if an object has mass—if it is made out of "stuff"—by using a scale. If the object weighs something, that is evidence that it has mass.

■ Demonstrating How to Use the Scales

1. **Hold up a spring scale.** Ask what will happen if you put a pencil into the bag clipped to the scale. [The pencil will pull down on the bag, which will pull down on the spring inside the scale.] Explain that the scale is designed so that the spring stretches out if something pulls on it. If the spring stretches, this is evidence that the pencil has mass.

2. **Set a pencil in the bag.** If you made a reference mark on an un-calibrated scale, tell the students it helps to watch how the spring moves in relation to the mark. Say that if the spring moves even a small amount, it means the object has mass. The more it moves, the more mass (and weight) the object has.

3. **Test another object.** Have students predict what will happen if you set a more massive object, like a key ring with several keys, in the bag. [It will pull down on the spring much more.] Try it. Ask if the object has mass. [Yes. It has more mass than the pencil.]

4. **Use an object that is too heavy to be measured by the scale.** Put the rock, or other heavy object, in the bag. Point out that it pulls the spring down as far as it can go. Ask if the object has mass. [Yes.]

If you are using calibrated spring scales and are going to ask the students to record the actual weight of the objects, carefully show them how to do this now. Show them how to record the weight of objects that are too heavy to be weighed on the scale by writing the maximum number on the scale with a plus sign (+) next to it.

5. Caution the students to treat the spring scales respectfully. Say that they should not stretch the spring scales out too vigorously. If they are using the red spring scales encased in clear tubes, they should not try to pull the spring out through the top or bottom of the tube.

■ Circulating to Stations

1. Review by asking, "How do you know the pencil is matter?" [It takes up space and has mass.]

2. Show how to record on the What's the Matter? sheet.

a. Go over the list in the left-hand column.

b. Show them where to record "yes" or "no" for the three questions. Model how to fill in the answers for an object: Does it have mass? Does it take up space? Is it matter?

3. Point out the blank spaces on the sheet. If students choose to test something that is not on the list, they can write it in a blank space, do the tests, and answer the three questions.

4. Pass out materials and begin. Tell the students they can go to stations in any order. Emphasize the importance of taking turns with the scales. Give each student a sheet and pencil and give each pair a scale. Assign them to their first station and have them begin.

■ Discussing Results

1. Class discussion. After your students have had enough time to test a variety of objects, collect the spring scales. (Students don't all need to get to all the stations.) Have them keep their **What's the Matter?** sheets and gather for a class discussion.

2. Ask if anything surprised them. Students may comment, for example, that the big Styrofoam object had relatively little mass and the tiny metal object had a lot of mass.

3. Ask, "What is one thing you tested that is matter?" For each response, ask how they know it is matter. [It has mass and takes up space.]

All the objects the students will be measuring at the learning stations should register on their scales. However, you may want to demonstrate that a paper clip or other light object might not register on the scale. Ask, "Does the paper clip have mass?" [Yes, you can feel it, but it doesn't have very much mass, and this scale might not be sensitive enough to measure it. A more sensitive scale would show that the paper clip has mass.]

You may want to ask students for their explanations for how a larger object could have less mass than a smaller object. This could serve as a very introductory glimpse into their ideas about the concept of density. However, thorough explanation of this complex topic is best saved for later grades.

4. Ask if they found any solids that are *not* matter? Any liquids?

5. Discuss the water in the balloon. If any students have classified water as "not matter," use the water in the balloon from the water station to do a demonstration of how it takes up space (the evidence is you can't put your fingers together) and has mass. (the evidence is it pulls down on the scale).

6. Tell the class that all solids and liquids are matter. Move the large Matter sign to above the Solids and Liquids signs in your class display. Ask what the students now know about all solids and liquids. [They are all matter and all have mass and take up space.]

■ Testing "Air"

1. Is air matter? Have the students hold a thumbs up or thumbs down to show whether they predict air is matter or not matter.

2. Show them an empty, deflated balloon. Put it on the calibrated spring scale and record how many grams it shows. Hold up the carbon dioxide-filled balloon you prepared. Tell them the balloon by itself has the same amount of mass as the other balloon, but this one has been filled.

3. Explain what's in the balloon. Tell the students that observing whether air has mass can be challenging. Explain that you have filled this balloon with one of the materials that air is made of. It is called carbon dioxide and now the class will see whether a scale will show if it has any mass.

4. Take the empty balloon off the scale and replace it with the CO_2-filled balloon. Ask, "Does the filled balloon have more mass than the empty balloon?" [Yes.] "Why?" [The stuff in the balloon has mass.]

5. Ask the students to use their fingers to test if the air in the room takes up space. They may report that no, it does not, because they can easily put their fingers together. Point out that because air can move around so easily, the air could be moving out of the way of their fingers as they close them.

6. Do the finger test on the balloon with CO_2 trapped inside it. Have a student do the finger test on the carbon dioxide-filled balloon, and ask, "Can you feel something between your fingers that takes up space?" Point out that the air in the balloon is trapped so it can't move out of the way.

Why not wait until Activity 5 to introduce gases? It's important to introduce the idea that air is matter briefly now because the issue usually arises during the "What's Not Matter?" discussion (below). However, don't worry if your students don't quite "get" that gases are matter yet. They will have an opportunity to gain deeper understanding in Activity 5.

This CO_2 balloon demonstration helps young students understand that air is matter, which is the goal here. The filled balloon has more mass than the empty balloon, and that's all young students need to know for now. They don't need to know that CO_2 has more mass per volume than the mixture of gases in the air we breathe, or consider density, or that the balloon's volume and the surrounding air pressure are also part of the story. For your own interest, and to find out why some popular demonstrations about the mass of air are flawed, please see Background for the Teacher.

7. **Is air matter?** [Yes, because it has mass and takes up space.] Put the balloon and the sign saying "air" under the Matter sign.

■ What's *Not* Matter?

1. **Ask if the students can think of things that are not matter—** things that don't have mass or take up space. Place the Not Matter sign off to the side of your display.

2. **Give some examples if necessary.** If your students are having trouble thinking of things that are not matter, share an example from the list below. Don't share too many, so they will have the opportunity to think of a few themselves. As students share examples, write each one on a piece of paper and post it under the Not Matter sign.

Some Examples of Things That Are Not Matter

- heat
- light
- electricity
- sound
- music
- creativity
- truth
- thoughts
- ideas
- feelings
- dreams
- singing
- shadow
- imaginary creatures
- words

3. **Ask the students to think of other things that are not matter to share in the next session.** Have them write their list on a sheet of paper, title it Not Matter, and add it to their journals.

■ Are People Matter?

1. **Ask for a volunteer to stand in front of the class.** *(Choose some-one who is not overweight.)* Ask the class if they think people, like the volunteer, are matter. Ask them to share their reasoning.

2. **Do people take up space?** Put your hands on either side of your volunteer's head. Pretend to push really hard. *(Contorted facial expres-sions are good.)* Tell the class you can't put your hands together be-cause there is something (very intelligent) between them that takes up space.

3. **Do people have mass?** Either lift the student or pretend to try, and tell the class that yes, you felt weight. Ask what would happen if the volunteer got on a scale. Confirm that the evidence shows that people do have mass.

4. **Ask again if people are matter.** [Yes, because they have mass and take up space.] Place the "people" sign under the Matter sign.

One entertaining teacher we know likes to tell students that if someone ever asks them (acted out by the teacher in a voice dripping with sympathy), "What's the matter?" they could now answer, in a humorously self-righteous tone (again acted out by teacher), "I am! I have mass and I take up space!" Students get a kick out of the joke (and it cements the definition besides!).

Your Solids and Liquids class display should now look something like this:

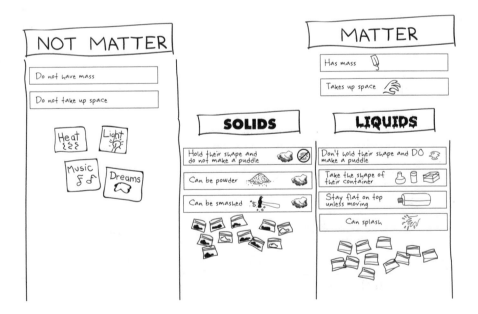

Name: _____

What's the Matter?

	Does it have mass? (Yes or No) 	Does it take up space? (Yes or No) 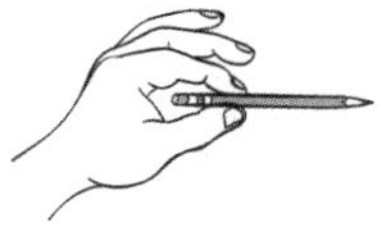	Is it matter? (Yes or No)
1. Wood			
2. Styrofoam			
3. Plastic			
4. Metal			
5. Water			
6. Paper			
7. Rock			
8. Glass			
9.			
10.			
11.			
12.			

Session 1:
How Do We Know Gases Exist?

Overview

In the first session of Activity 5, students begin by sharing their new suggestions of things that are "not matter." Any that are confirmed by the class discussion are added to the class display.

Next, the students are challenged to come up with evidence that air (gases) do exist. The class discusses the evidence, gas is established as a third kind of matter, and posted on the class display. Then, through a brief demonstration with peppermint extract, the students learn some attributes of gases: They are often invisible, they can spread around in the air, and they can sometimes be detected by our sense of smell.

Session 1 is shorter and less active than most sessions in the *Matter* unit. However, it gives students an opportunity to reflect on the challenging subject of whether gases are really matter, and prepares them for the interactive investigations in Session 2.

Most teachers find that Activity 5 is best presented in two separate class sessions, as outlined in these pages. However, if you have older students and a generous block of time, you might plan to present both Sessions 1 and 2 in one long session.

Learning Objectives for Activity 5, Session 1

Students will:

- learn that gases are observable
- learn that smells are evidence that gases spread in the air
- support arguments with evidence

■ What You Need

- ❏ a chalkboard or piece of chart paper
- ❏ the Solids and Liquids display from previous sessions
- ❏ a large sign that says Gases
- ❏ several sentence strips
- ❏ 1 black wide-tip marker
- ❏ a clear container or cup
- ❏ about 1 tablespoon of peppermint extract in a closed bottle
- ❏ a small amount of water (1 tablespoon is enough)
- ❏ a pan and hot plate OR a shallow container or plate

Heating peppermint extract on a hot plate makes its scent spread throughout the classroom most effectively. Keep the extract in a bottle until the activity, then warm it in a pan or shallow container that can withstand heat. If you won't be using a hot plate, simply pour some extract into a shallow container or plate to release the scent.

■ Getting Ready

1. **Make a large sign that says Gases.** (You may want to write the word in "gassy," cloud-like letters.)

2. **Write a sentence strip sign** that reads "spread in the air."

3. **Have your peppermint extract handy,** but out of sight.

   Optional: **Set up a hot plate** to warm the extract.

4. **Fill a clear container or cup partially with water.**

■ Introducing Gases

1. **Add examples of things that are not matter.** Review any examples of Not Matter already posted on the class display, such as thoughts, dreams, or shadows. Ask, "Who has thought of more things that are not matter?" If everyone agrees that something does not have mass and does not take up space, write it on a sentence strip and add it to the display.

2. **Remind the students that all solids and liquids are matter.** Hold up the container partially filled with water and ask the students to show hand signals for solid (a fist) or liquid (a slowly waving hand). Review that the container is a solid, the water is a liquid, and both are matter.

3. **Ask what else is in the container?** Tell the class there is something else in the container. Ask what it might be. [Air or gases.] Tell them air is a mixture of *gases.*

■ Thinking of Evidence that Gases Exist

1. **Pretend for the moment that you don't think gases exist.** Say, "I can see liquids and solids, but when I try to look at air, I don't see air, I see the chalkboard. If I try to grab air, I get nothing."

2. **Challenge the students think of examples of evidence that gases/air exist or are real.** As students suggest evidence (ways to show) that air (or any gas) exists, have everyone try out their suggestions. For example, if a student says, "if you wave your hand you can feel air," ask everyone to try waving their hands and seeing if they feel evidence that gas exists.

3. **List their suggestions.** Write their ideas on chart paper or the board where you can refer to them in the next session. Be sure to let them know that their evidence is convincing, and that you do think that gases exist.

Sample List of Evidence That Gases Exist

- If gas is trapped in a balloon, you can feel that it takes up space.

- If you wave your hand you can feel gas.

- You can feel gas when you blow on your hand.

- You can hear and feel gas when you breathe in and out.

- If you blow upwards, the gas makes your hair move.

- If you fan yourself, you can feel gas on your skin.

- If you run fast, you can feel gas on your face.

- The wind is gas that is moving, and you can see it moving leaves.

- Sometimes you can smell gas.

- You can hear the hissing sound of gas escaping from a tire or balloon.

- On cold mornings you can see your breath. *(See note)*

---

4. **Gas exists.** Thank the students for their evidence. Acknowledge that, while it may be invisible, gas does exist.

When first introducing the term gas it's often useful to use it in conjunction with the word air (or "airs" as children often call it). Later, as students gain a better understanding of gases, shift to using just the term gas.

The last idea on the list (On cold mornings you can see your breath) is acceptable as evidence for gas, but it's actually not gas one is seeing. Warm air holds more water gas than cold air. When the warm air from your mouth comes in contact with cold air, some of the water gas condenses into liquid (not gas) water droplets. The fog you see is not gas, but condensed liquid water droplets suspended in the air. This explanation is probably beyond most of your students at this stage, so it's probably best to simply accept "seeing your breath" as evidence for gas.

■ Gases Are a Third Kind of Matter

1. **Ask, "Is gas matter?"** Remind the class of the gas-filled balloon in Activity 4 that weighed more than the empty balloon. That was evidence that gas has mass. By squeezing the balloon with her fingers, a person could feel that gas takes up space. [Gas is matter because it has mass and takes up space.]

2. **Add the large Gases sign to the class display.** Say that gases are a third kind of matter. By the end of the unit, your class display will look something like the drawing below.

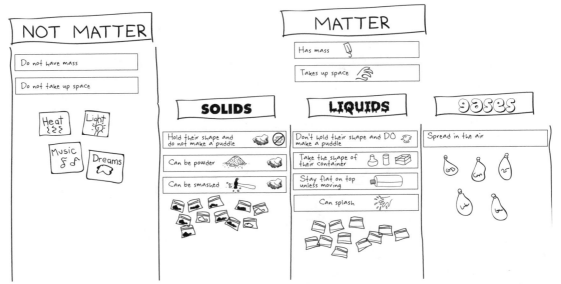

3. **Give the students a hand signal for gases:** a flat hand with wiggling fingers. Hold up the container of water again and have them show the hand signals for solid (fist), liquid (slowly waving), and gas (wiggling fingers).

■ Peppermint Demonstration: Gases Can Spread Out and Move Around

1. **Our noses are gas-detectors.** If no one has mentioned it, say that sometimes we can smell gases. Tell the class we all have "gas-detectors" built into our faces—our noses. We can't smell all gases, but we can smell many of them.

2. **Rephrase examples.** Ask students to share examples of smells they have experienced, then rephrase each of their examples in terms of gases. For example:
   > Student: I smelled cookies.
   > Teacher: So that means gas from the cookies spread through the air and got sucked into your gas-detecting nose. You were smelling cookie gas!

3. **Challenge the students to detect a gas.** Tell them you're going to let a pleasant-smelling gas into the air in the room. (Don't tell them what it is.) It's their job to wait with their eyes closed, and raise their hands if they smell this gas. It's important that they not raise their hands unless they are sure they smell it.

4. **Open peppermint extract.** When their eyes are closed, open the bottle of peppermint extract and heat the extract in a pan on a hot plate or pour it into a shallow container or plate.

5. **Ask, "Who smells peppermint?"** When many hands are up, tell the students to open their eyes and look around the room to see whose hands are up. Show them the liquid peppermint in the dish.

6. **How did the peppermint smell get to your noses?** Ask, "Did the peppermint *liquid* spread in the air to get to your nose?" [No.] Ask, "How did the smell from the peppermint liquid get to people's noses?" [The smell was evidence that gas came from the peppermint extract in the pan or plate and moved through the air.]

7. **Gases move and spread around more easily than solids or liquids do.** Explain that gases are different from solids and liquids, because they don't stay the same size. A solid rock or a liquid in a bottle stay the same size. But gases can spread out and move around, mixing into the other gases of the air.

8. **Post the sentence strip *"spread in the air"* under the Gases sign.** Invite students to suggest other statements about gases. If everyone agrees, write them on sentence strips and post them on the display. Say that in the next class session, they will get to learn more about gases.

This activity works dramatically better if the extract is warmed on a hot plate.

Please see the Background for the Teacher section about how gases move. The more complete story is beyond the scope of the unit, but may be helpful for you in responding to some student questions.

One teacher gave her students the example of opening a lunch box and smelling the gases that come from the different solid foods and mix with the air.

Another teacher commented, "Of course they wanted to discuss their own 'personal' gas—and even though we didn't discuss it much, it helped them understand the concept!"

For another demonstration, use a turkey baster to suck up a scent, such as the peppermint gas, from just above the extract, and transport it to near a student's nose in another part of the room.

Session 2: Investigating Gases

Overview

In the second session of Activity 5, student pairs circulate to a variety of "gas stations." Each station explores gases, such as smelling gases, "pushing" gases with devices such as fans and blowers, making gas by mixing vinegar and baking soda, comparing light and heavy gases in balloons, and moving gases around using big bags.

After the learning station activities, the class discusses the evidence they have gathered that gases exist. They also review the evidence that gas is matter because it has mass and takes up space. They are then challenged to classify a variety of objects as solids, liquids, gases, or a combination. At the end of the session, the students add drawings representing various gases to their journals.

One teacher said, "Interest and excitement levels were high. Either I'm getting used to the prep time, or it's worth it. Preparation for the station activities seemed easier and smoother."

Learning Objectives for Activity 5, Session 2

Students will:

- learn that gases are matter because they take up space and have mass
- gather evidence about the properties of gases through a variety of tests
- learn that all solids, liquids, and gases are matter
- know that an object can be made of a combination of solids, liquids, and gases
- support arguments with evidence
- follow directions, work cooperatively in groups, and record data

---

■ What You Need

For the class
- ❑ Solids, Liquids, and Gases display from previous sessions
- ❑ List of Evidence That Gases Exist generated in Activity 5, Session 1
- ❑ 50 address labels or a roll of masking tape for labeling the bags
- ❑ 4 black fine-tip permanent markers
- ❑ 50 pushpins
- ❑ a few solid, liquid, and gas items from previous sessions, including a clear container half-full of water
- ❑ 50 sealable, clear snack-size bags
- ❑ copy of photo of helium balloon in a vacuum (page 87)
- ❑ *(optional)* an electric fan or hair dryer

For each student
- ❑ journal
- ❑ one copy of **Gases** student sheet (page 86)

For Learning Station #1: Masses of Gases
- ❑ 1 tethered helium balloon
- ❑ 1 balloon filled with air
- ❑ 1 balloon filled with carbon dioxide gas (made by mixing vinegar and baking soda)

For Learning Station #2: Moving Gases
- ❑ 2 large plastic trash bags
- ❑ about 5 lightweight plastic grocery shopping bags

If you use squeeze bottles, students can squirt smelly gas samples into their bags.

For Learning Station #3: Smelly Gases
- ❑ a few drops of 4 scents—extracts, flavorings, or perfumes
- ❑ 4 film canisters with lids (or squeeze bottles)

For Learning Station #4: Gas Pushers
- ❑ 1 trash can
- ❑ scratch paper made into a few paper fans
- ❑ 32 straws (or enough for each student to have one)
- ❑ 1 air pump (balloon, sports ball, or bicycle)
- ❑ 1 turkey baster
- ❑ a few Styrofoam packing peanuts
- ❑ a few small pieces of paper
- ❑ a few paper clips
- ❑ a few coins or small washers
- ❑ *(optional)* 1 syringe without a needle

For Learning Station #5: Making Gas (2 stations)

- ❏ one 16-oz. bottle of white vinegar
- ❏ 4 squirt bottles for vinegar (to prevent spilling)
- ❏ 1 permanent marker
- ❏ 1 16-oz. box of baking soda
- ❏ 2 cups
- ❏ 4 copies of Making Gas procedure sign (page 85)
- ❏ 12 plastic vials (no lids)
- ❏ 4 graduated cylinders (for measuring 15 ml)
- ❏ 4 plastic teaspoons
- ❏ about 20 balloons (see note, below)
- ❏ 2 funnels
- ❏ 2 sponges
- ❏ *(optional)* 2 dishtubs or bowls for rinsing vials

You'll need one balloon for each pair of students, plus two balloons for the Light and Heavy Gases station. We suggest getting a package of balloons that inflate to about 4–6 inches in diameter. Test to make sure they are easy to inflate. Don't use water balloons.

■ Getting Ready

Before the Day of the Activity

1. **Decide how many learning stations you will need to set up.**
   Students will circulate among the learning stations. There are five different activities. We recommend that you set up two stations of the fifth activity, Making Gas, because it takes longer than the others and the station may get crowded. This will make a total of six learning stations, each accommodating about four students at a time. If you have more than 24 students, duplicate one or two of the other stations, such as Gas Pushers or Smelly Gases.

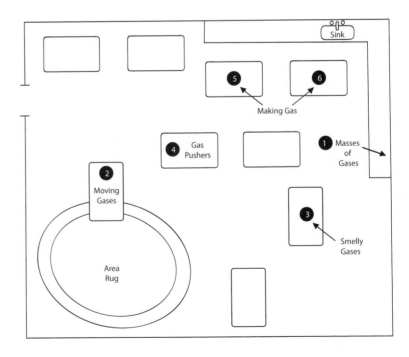

2. **Decide how you will arrange the room.** Think over how you will set up the learning stations on tables or counters or at students' desks. Have enough space between them for students to move about as freely as possible. Plan to set up the Moving Gases station where there is space for students to wave bags through the air to collect gases. If possible, put the Making Gas stations near a sink.

3. **Recruit adult volunteers.** Arrange for two adult volunteers to help at the two Making Gas learning stations. The volunteers will supervise students as they mix baking soda and vinegar in a balloon. They will help with tying the balloons and keeping the stations clean.

4. **Helium balloon.** Ask one of the volunteers, or someone else, to bring in a helium balloon on the day of the activity for the Light and Heavy Gases station.

5. **Make copies.** Make a copy of the **Gases** student sheet (page 86) for each student. Make two copies of the procedure sign for each Making Gases Learning Station (page 85). Make one copy of the photo of a helium balloon in a vacuum (page 87).

On the Day of the Activity

1. **Set up the learning stations.**

   a. **Masses of Gases Station:** Fill one balloon by blowing air into it. Use some materials from the Making Gas Station to fill another balloon with carbon dioxide gas. (See the Making Gas procedure sign (page 85) for directions.) Tie the string of the helium balloon to a chair near the other balloons. Label the balloons: air, carbon dioxide, helium.

   b. **Moving Gases Station:** Set out 2 large plastic trash bags and about 5 lightweight plastic grocery shopping bags. (Students swing bags to capture gases, so allow space near this station.)

   c. **Smelly Gases Station:** Add a very small amount (a drop or less) of one scent to each of the canisters. Put the lids on.

   d. **Gas Pushers Station:** Set out several paper fans, a turkey baster, enough straws for each student to have one, an air pump, and *(optional)* a syringe without a needle. Set out objects for students to try to move with the gas pushers—a few Styrofoam packing peanuts, small pieces of paper, paper clips, and coins or small washers. Set a trash can near the station for used straws.

e. **For each Making Gas Station:** Fill two squirt bottles with vinegar and label them. Fill a cup about half-full of baking soda. Set out:
- 2 copies of the Making Gas procedure sign
- the vinegar in squirt bottles
- the cup of baking soda
- 6 plastic vials
- 2 graduated cylinders
- 2 plastic teaspoons
- about 10 balloons
- 1 funnel
- 1 sponge

2. **Set up a Labeling Station:** The students will need a place to label their bags, located near the class display, if possible. Set out masking tape or address labels, four permanent markers, and a container of pushpins. (Hold on to the sandwich bags to hand out just before the activity.) If you have students who are pre-writers, write, "CO_2," "Smelly Gas," and "Air" on some labels and place the labels at the Labeling Station.

3. **A few materials for the class discussion.** Have handy a few solid, liquid, and gas items from previous sessions, including the container half full of water. *Optional:* Bring in an electric fan or hair dryer to use as you discuss the Gas Pushers station.

■ Gas Stations

Do they sell "gases" at gas stations? *When you say the word "gas," children may be confused because they think of gas stations. We usually don't see what is being pumped at a gas station, so we don't see that the gasoline is a liquid. You can point out that the term "gas station" is an abbreviation for gasoline station and that it's actually liquid gasoline that comes through the pump, not a gas. Of course, if you can smell it, and you usually can, you are smelling the gas coming from the evaporating liquid gasoline.*

1. **Gases are a third kind of matter.** Briefly review the three kinds of matter posted on the class display.

2. **Exploring more evidence of gases.** Tell the students they are going to do activities with gases at learning stations around the room. The activities will give them more evidence that gases do exist and that gases are matter.

3. **Assign partners.** Explain that the students will be working with a partner at the stations. They should cooperate with their partners and discuss what they discover. Explain your expectations for good classroom behavior.

4. **Collecting gases.** Point out that at Stations 3 (Smelly Gases), 4 (Gas Pushers), and 5 (Making Gas), each pair of students will collect one sample of gas in a bag to put on the class display. At Stations 1 and 2 they won't collect gases.

5. Point out the labels, pens, and pushpins at the Labeling Station. When partners collect a bag of gas, they will make a label with the name of the gas, put it on their bag, and pin the bag on the class display under Gas. Ask them to pin the very top of the bag so it doesn't pop.

6. Explain the procedure at each station.

Station #1: Masses of Gases
There are three balloons, each filled with a different kind of gas. Try to figure out which has the most mass (is heaviest) and which has the least mass (is least heavy). **Don't collect gases for the class display.**

Station #2: Moving Gases
Ask, "Can you fill these empty plastic bags with gas/air?" Tell the students they will only fill the bags temporarily, and will not be tying them off. Caution them to be careful not to get too wild at this station.

Station #3: Smelly Gases
Open the canisters one at a time, smell them, and put the lids back on. Pick **one** smelly gas to collect in a snack-size bag. Hold the bag over the open canister for a few seconds. Seal the bag. Go to the Labeling Station and label the bag. *Note: If you have put the scents in squeeze bottles, tell students to use them to squirt the smelly gases into their bags.*

Station #4: Gas Pushers

a. Try different gas pushers: air pump, syringe, turkey baster, paper fan, and straw. Tell the students the straws are not for sharing. When finished, they should throw away their straw.

b. Which gas pusher is best for moving things? Use each gas pusher to see if gas can move Styrofoam, paper, paper clips, and coins or washers. If it moves the Styrofoam and paper, it's a pretty good gas pusher. If it moves paper clips, it's even better. If it moves washers or coins—wow!

c. Collect a gas sample. Use one of the gas pushers to push air/gas into a bag, seal it, label it *air,* and pin it in the Gas section of the class display.

Emphasize that students will be collecting only gases. They shouldn't include any liquids or solids.

Show younger students the pre-written labels they will use.

Station #5: Making Gas

a. **Two identical stations.** Explain that there are two Making Gas stations. They are exactly the same, so students should go to only one of them. An adult volunteer will be at each Making Gas station to help if needed.

b. **Demonstrate the procedure.** To avoid spoiling the surprise, don't actually do Step 4. Tell the students there is a sign at the station to remind them of all these steps in case they forget.

1. Measure 15 ml. of vinegar using the graduated cylinder and pour it into a vial. (*Optional:* measure 1 tablespoon) Make sure students understand what is meant by "ml" and vial.

2. Measure 1 level teaspoonful of baking soda and pour it into a balloon using a funnel.

3. Stretch the opening of the balloon over the mouth of the vial without spilling the baking soda into the vial.

4. Lift the end of the balloon so the baking soda falls into the vial, and watch what happens.

5. Ask for an adult to help take the balloon off (without deflating it) and tie it.

6. Label the balloon CO_2 and place it in the gas section of the class display.

7. **Assign pairs to their first station.** Give each student pair three sandwich bags and remind them to collect gases at Stations 3, 4, and 5. Tell them that after their first station, they can go to the other stations in any order. They should go to less crowded stations as much as possible.

■ Discussing Gas Stations

1. **Gather for a class discussion.** When pairs have finished all five activities, gather the class away from the stations, where they can see the class display. Compliment the class on their collection of gas samples on the display.

2. Refer to the List of Evidence That Gas Exists from the last session. Ask the students to think of evidence from the Learning Stations that gas exists. Hold off on discussing Station #1 (Masses of Gases) until Step 3. For each other station ask, "What is your evidence that gas exists, even though you can't see it?" Add this evidence to the list the students generated in the previous class session. For example:

Moving Gas: We captured the air in the bag, and the bag swelled up. This is evidence that air exists and takes up space. (It's matter.)

Smelly gases: What we smell is the gas coming out of the container. The smell is evidence that there is a gas.

Gas Pushers: Gas made objects move. Also, we can feel the gas moving. That is evidence that gas can move things, is real, and has mass.

Making Gas: When we mixed vinegar and baking soda it made carbon dioxide gas. It filled out the balloon, which is evidence that it is "real."

> *Sample List of Evidence That Gases Exist*
> • If gas is trapped in a balloon, you can feel that it takes up space.
> • If you wave your hand you can feel gas.
> • You can feel gas when you blow on your hand.
> • You can hear and feel gas when you breathe in and out.
> • If you blow upwards, the gas makes your hair move.
> • If you fan yourself, you can feel gas on your skin.
> • If you run fast, you can feel gas on your face.
> • The wind is gas that is moving, and you can see it moving leaves.
> • Sometimes you can smell gas.
> • You can hear the hissing sound of gas escaping from a tire or balloon.
> • On cold mornings you can see your breath.

3. **Use Station #1 to emphasize the evidence that gas is matter.** Refer to the definitions of matter on the class display. [Take up space. Have mass.]

4. **Does gas take up space?** As in Activity 4, try to put your fingers together around one of the gas-filled balloons. Ask why you can't put your fingers together? [Because there is gas inside the balloon that is taking up space.] Tell the students that **all gases take up space.**

5. **Does gas have mass?** Ask, "Which of the gases (helium, carbon dioxide, or air) in the balloons had the most mass?" [Carbon dioxide.] "The least mass?" [Helium.]

6. **Does helium gas have mass?** Tell the students that helium does have mass, but it has less mass than air. That is why it floats above the other air in the room.

7. **Hold up the picture of the helium balloon in a vacuum.** Say that someone put a helium balloon in the container, then pumped out all the air around it. Why does the helium balloon sink to the bottom of the container? [This is evidence that helium gas has mass. It falls because of the pull of gravity between its mass and Earth. You could measure its weight (mass) on a scale, as long as there is no other gas in the container.]

8. **Ask again, "Is gas matter?"** [Yes! Even gases with only a little mass are matter.] "How do you know?" "What is your evidence?" [Gas is matter because it has mass and takes up space.]

■ More Challenging Substances

1. **Three hand signals.** Ask the students to show you the hand signals for solid (a fist) and liquid (a slowly waving hand). Remind them of the hand signal for gas (wiggling fingers).

2. **Hold up a few items.** Ask the students to use hand signals to indicate whether they think they are solid, liquid, gas, or a combination. Be sure to include a few that involve all three (such as a partially-filled container of liquid).

■ Classifying People

1. **People are matter.** Remind the students that in the previous session they agreed that people are matter, because they have mass and take up space.

2. **Which kind(s) of matter are we?** Ask the class to use their hand signals to show whether they think people are made of solid, liquid, or gas. Say they can show more than one hand signal.

3. **Share ideas and review definitions.** Ask the students showing each different hand signal to explain their thinking. As they explain, review what makes something a solid, liquid, or gas. For example:

   a. **People are solid.** A whole person could be called a solid because we hold our shape and don't turn into a puddle! Also, many parts of us are solid, for example, our teeth and our hair.

   b. **People are made of liquids too.** Our blood is a liquid. [It can make a puddle.] Have the students feel the saliva in their mouths with their tongues.

   c. **People are made of gas too.** Tell the students to take in a breath of air. Tell them that they are also gas, because the gases they breathe in are spread in the blood throughout their bodies.

4. **People are made of all three kinds of matter.** Is a person completely solid? Completely liquid? Completely gas? [No, we are made of three kinds of matter.] Say that this is true of all living things: bananas, pine trees, mosquitoes, and bears are all made of solids, liquids, and gases. Many objects are made of more than one of these three kinds of matter.

5. **Scientists study matter in the Universe.** Tell the class that what they have learned about matter is important, since everything in the Universe is made of it! Tell them their understanding of solids, liquids, and gases will help them in later grades as they continue to study matter.

■ Drawing Gas

1. **Introduce the Gases student sheet.** Say that to complete their Matter journals, they will draw gases on a student sheet. Ask for their ideas on how they might draw gases, even though they are invisible.

2. **Share some examples.** Have a few students share where there were gases at the learning stations. Show one or two examples of how to draw the gases on the board. For example, draw a film canister with wavy lines to represent the smelly gas coming out, or draw a gas pusher, such as a straw, with wiggly lines coming out the end. Show how to label the wavy lines "gas."

3. **Pass out student sheets.** Allow a few minutes for the students to record gases on the sheet. They can draw other gases besides the ones at the Learning Stations. If possible, allow time later for students to share with their classmates some of the gases they recorded.

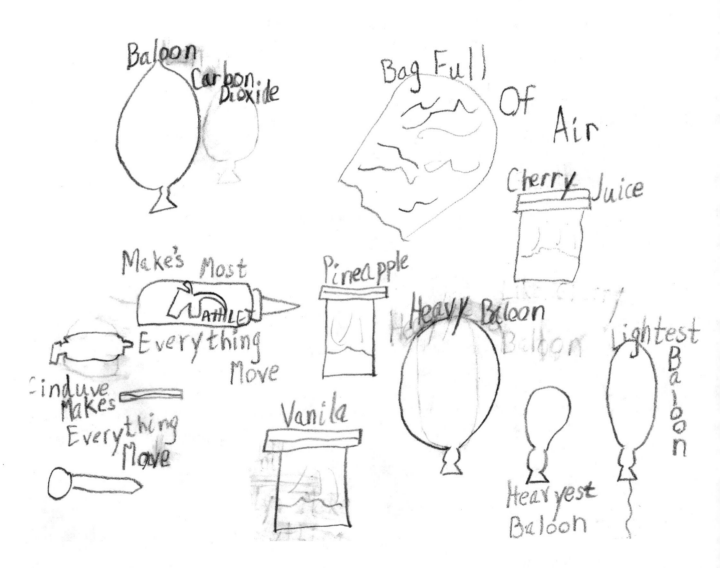

Sample of Student Work

Making Gas Procedure Sign

1. Measure 15 ml. of vinegar and pour it in a vial.

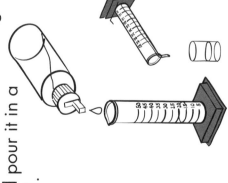

2. Measure 1 level spoonful of baking soda and pour it in a balloon using a funnel.

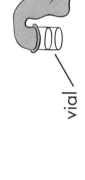

3. Stretch the opening of the balloon over the mouth of the vial without spilling the baking soda into the vial.

vial

4. Lift the end of the balloon so the baking soda falls into the vial, and watch what happens.

5. Ask for an adult to help take the balloon off (without deflating it), tie it and label it "carbon dioxide."

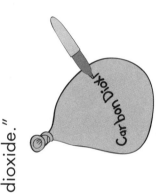

Carbon Dioxide

6. Place it in the gas section of the class display.

GASES

Carbon Dioxide

gases

You can tell something is a gas because:
• It spreads in the air. (It doesn't hold its shape.)

Many gases are invisible, but you can draw where they came from. You can draw wiggly lines to show gases.

Draw and label gases here:

The following is intended to provide you, the teacher, with information that should be helpful in responding to student questions. It is not meant to be read out loud to or duplicated for students. It contains information for beyond the grasp of this age group, but which may be helpful for the teacher.

What Is Matter?

The discussion of what constitutes matter can expand in lots of intriguing directions. Students may say things like, "You have to be able to touch it or see it for it to be matter." But there is more to matter than that! Some materials and/or phenomena provoke controversy and discussion. This is good, for it helps students hone their own conceptual understanding, and it deepens awareness of scientific complexity and the fact that science is a constant questioning process.

For example, as students learn in this unit, air is matter, and is made up of different gases. Gases have mass and take up space. But consider wind. Wind is a current of air, and air is matter. Air has mass, but would you say that a current of air has mass? Is wind matter?

A star is a mixture of fiery gases. Our Sun is a star that is composed mostly of hydrogen and helium gases. So the Sun is matter. But the life-giving energy we soak in from the Sun, which comes to us as light, is not matter. It is fine, as a reflection of science inquiry, to ponder these complex matters without deeming one particular answer right or wrong. Is a beam of sunlight matter? If we think of a beam of sunlight as the air molecules that have been heated up by the Sun, then matter is involved in that beam, so the question is worth discussing if a student brings it up!

The basic definition of matter is not entirely clearcut even for scientists. Matter must have some sort of substance. How that substance is measured, observed, or detected is up for debate. Stay open to possibilities.

What Are Solids, Liquids, and Gases?

Solids, liquids, and *gases* are terms used to group matter according to definite mass, volume, and shape. Volume is a measurement of how much space a substance occupies. Mass is a measurement of how much matter is in a substance.

Here is one common set of definitions for solids, liquids, and gases:
— Solids have definite mass, volume, and shape.
— Liquids have definite mass and volume, but not shape.
— Gases have definite mass, but not volume or shape.

For young children, these can be loosely translated into:
— *Definite shape* keeps (or holds) its shape.
— *Definite volume* keeps its size and amount.
— *Definite mass* keeps its same weight.

If your students bring up these ideas, by all means discuss them and add them to the list. But don't impose these ideas if they don't make sense to your students. For example, it is common for younger children to find it difficult developmentally to understand that a quantity of liquid poured from a tall and thin container into a short and fat one is still the same amount of liquid. If they cannot understand this, they won't understand the *definite volume* definition when applied to liquids.

Mass and Weight

Mass is an intrinsic property of an object that quantifies how much matter is in it. This quantity is the same no matter where the object is—on Earth, on the Moon, or in outer space.

Weight, by contrast, is a force. Scientifically, it is the force of gravity on the object. For instance, on Earth weight is the force between the Earth and the object that pulls downward on the object. On the Moon it would be the force between the object and the Moon. Since gravity is less on the Moon than on Earth, the object's weight is less on the Moon than on the Earth. (But remember, its mass is the same no matter where it is.)

On Earth, an object's weight is directly proportional to its mass, and so the two words are often used interchangeably. "The boulder weighs a lot" means that it has a lot of mass.

The common definition of matter is that it has *mass* and takes up space. As noted in this guide, mass is a harder concept than weight for young students to grasp, but they can use the term and begin to get an intuitive sense of what mass is. As the unit recommends, students should start using the word *mass.*

More on Weight and Mass

Weight is the force of gravity on an object. This is an accurate scientific definition. In speech, however, the word *weight* is used to mean the force that a scale measures. These are almost the same in most cases, with some important exceptions. On Earth the measurement made by a scale is not exactly the same as the force of gravity because of buoyant forces. We are surrounded by air, which exerts a slight upward force on everything. This makes the measurement of a scale slightly less than the force of gravity. In water this difference is much larger and more noticeable. You feel "weightless" in water.

When considering the weight of gases, this becomes a bigger issue. Air is pulled by the force of gravity, but it measures zero force on a scale. A helium balloon is pulled by the force of gravity, but it measures a negative force on a scale! Objects in orbit (consider a space station and the astronauts inside) are subject to the force of gravity. That is what keeps them in orbit. So by the scientific definition, they have weight. But, on a scale, they are apparently weightless because a scale will show zero force. Scientifically speaking, they are not weightless, but the word has entered speech because they *feel* as if they are. This is only relevant to this unit because students need to realize that objects in orbit described as being weightless have mass and are still matter.

Mass is more scientifically universal than weight because it does not depend on the gravitational conditions of the Earth. Mass is determined by how much force it takes to change the velocity of an object within a certain time. It takes more force to get a cement truck to go from zero to 60 mph in 30 seconds than it does to get a small car to go from zero to 60 mph in 30 seconds—the cement truck has greater mass than the car. The amount of force needed to change the cement truck's velocity would be the same everywhere—on Earth, on the Moon where it would *weigh* less, underwater where a scale would show it lighter in weight, and in orbit where a scale would show it weighing nothing at all. On Earth, a ping-pong ball and a golf ball have different weights, but in orbit a scale would show them weighing nothing. Still, a speeding golf ball would hurt more if it hit you on the head than a speeding ping-pong ball would—on Earth, on the Moon, anywhere. The force it takes for your head to stop the ball (to change the ball's velocity) depends on the ball's mass, not its weight. **However, this unit does not require that students understand about mass as it relates to the force**

required to change an object's velocity. It is accurate enough to say, "We can tell about an object's mass by seeing how heavy it is." To students who ask how weight and mass are different you can say, "The weight of this object would be different on other planets, but its mass is something that stays the same."

Regarding the Weighing of Air

In Activity 4 there is a demonstration of the mass of air using a spring scale and a balloon filled with carbon dioxide. Using carbon dioxide allows you to genuinely demonstrate the weight of a gas, even though the gas is not the same as the mixture of gases in the air.

There is a well-known, but incorrect, demonstration that purports to show that air has weight. A beam-balance is constructed using a meter stick. Deflated balloons are attached to the ends, and the balance is adjusted. One balloon is then inflated, and that end of the balance beam sags downward, supposedly because the air in the balloon weighs it down. Actually, this experiment would not work if air did not have mass. The reason it works is not simply because air has weight. It is much more indirect than that.

See page 98 for another approach to showing that air has mass.

Ordinarily there is no downward force on a packet of air when that packet is immersed in more air. For example, think about a small bit of air that's right in front of you. It experiences no net upward or downward force. If you put a small plastic bag around that small bit of air, there's still no upward or downward force on the air inside. But when you put a balloon around that packet of air, it is under slight pressure because of the inward pull of the stretched rubber. That is, the stretched rubber of the balloon compresses the air slightly so that it is a little bit more dense than the surrounding air. It is thus slightly less buoyant and it causes the balloon to drop a bit.

If you did this experiment with two zip-locking bags, balancing one bag mostly full of air against one that is flattened, the demonstration would not work because the air-filled bag would not stretch and compress the air. In fact, even with a balloon you have to blow it up so that the rubber is stretched very tightly in order to detect a difference using the beam-balance. Again, you are not directly detecting that air has weight—you are detecting that compressed air is less buoyant than the surrounding air. Direct weighing of air works only if performed in a vacuum environment (say, on the Moon's surface).

States of Matter from a Molecular/Energy Perspective

In future grades your students will be exposed to more sophisticated models of matter, involving atoms, molecules, and energy. A key concept is that various states of matter differ in the amount of energy they possess. (In the GEMS *Dry Ice Investigations* unit for Grades 6–8 students explore this idea.)

Solids are formed when the attraction between individual molecules is greater than the energy causing them to move apart. Because the molecules are locked in position near each other, a solid has a defined shape and volume. Sometimes the bonds that hold molecules in place are flexible, as in rubber, or breakable, as in chalk, but these materials are still solid.

Liquids are formed when the energy in a system is increased and the rigid structure of the solid state is broken down. In a liquid, molecules can move past one another and bump into other molecules, but they remain relatively close to each other. As a result, a liquid can "flow" to take the shape of the container it is in but it cannot be easily compressed. Thus liquids have an undefined shape, but a defined volume.

Gases are formed when the energy in the system exceeds the attraction between molecules. In the gaseous state, molecules move quickly and are free to move in any direction, spreading out anywhere within their container. A gas thus expands to fill its container and has a low density (relative to a liquid or solid.) Because individual molecules are widely separated and can move around easily in the gaseous state, gases can be compressed easily and they have an undefined shape.

Liquids and Gases

What is the difference between a liquid and a gas? Both are fluids; both can flow. Gases are usually less dense than liquids, although some gases under high pressure can be more dense than some liquids. The main difference is that gases are a different phase of matter. *A gas can be made to condense into a liquid form, and a liquid can be made to evaporate into gas.*

It is often said that one important difference between liquids and gases is that gases in a container *expand to fill* the whole container and liquids do not. This is true for a container that is otherwise completely empty—a vacuum. It does not work for containers that are pre-filled with air, as our environment and most containers are.

In an air-filled room, dense gases act much like liquids; they can be poured into a cup or bowl, or spilled out onto a tabletop. If you release some carbon dioxide out of a CO_2 fire extinguisher, it will not instantly expand to fill the room. Instead it will pour downwards like an invisible fluid and form a pool on the floor. After some time it will mix with the rest of the air in the room, but *mixing* is very different from expanding to fill. Only in the world of the physicist, where *empty container* implies a vacuum, does the rule about gases expanding to fill the entire space work properly.

Misconceptions Your Students May Have About Matter

- Younger children often don't distinguish between an object (a spoon, for example) and the substance or material it is made of (metal, plastic, or wood).

- Children sometimes think that objects are made up of a combination of other objects, not of substances. For example, they may describe a bicycle as being composed of pedals, wheels, and other parts, but the actual substances the parts are made from are not perceived or understood.

- Powdery substances (such as chalk dust) are often mistakenly considered liquids because they can be poured. They may also sometimes be seen as consisting of a different kind of matter than other non-powdered materials or objects made of the same substance (such as a piece of chalk).

- As children begin to learn distinctions between states of matter, they may mistakenly assume that materials can exhibit properties of only one state of matter.

- Some children may think that everything that exists is matter, including heat, light, electricity, and shadows.

- Some children in elementary grades attribute weight only to heavy objects, while lighter objects are mistakenly considered to be weightless.

- Children sometimes assume that solids are always heavier than liquids, and liquids always heavier than gases. Gases are sometimes mistakenly considered to be weightless.

- Interestingly, in children's early conceptions, weight as a property of objects is often mistakenly overlapped or associated with the property of hardness.

- In general, the category of gases poses particular difficulties for students. Some misconceptions about gases include:

  — Children may mistakenly think that air cannot be captured or transported.

  —Students may think gases are not matter because most gases are invisible.

  — Students may think gases do not have mass.

  — Students may think air is "good" (because it's for breathing) but otherwise and in general gas is "bad" (because it can be poisonous, dangerous, and flammable).

  — Smell is often associated with the actual physical action of smelling, but may not be understood as evidence for the presence of gases.

  — Gases are often confused with automobile gasoline, but gasoline is a liquid, not a gas!

How Do Scientists Classify Challenging Substances?

Many substances are challenging to classify because they combine attributes of, or are mixtures of, a liquid and a gas, or a solid and a liquid, or all three. A *suspension* involves particles of one substance suspended in another. For example, sand in water is a suspension. Dust in air is also a suspension. In both cases, the solid (sand or dust) will settle to the bottom if left undisturbed.

A *colloid* is also a mixture of small particles of a substance dispersed in another in which it does not dissolve, but the particles are smaller in a colloid than in a suspension. Fog (or mist) and smoke are examples of colloids, as are toothpaste, Glook, shaving cream, and milk. Further subdivisions are also possible, including *foams, mists,* and *emulsions.* An **emulsion** is a colloid consisting of tiny particles of one liquid dispersed in another liquid—for example, salad dressing.

Sand, even though it "flows," is a solid. The reason it is classified this way becomes clear when you consider that when scientists classify a material such as sand, they are referring to *each grain* rather than a pile of it. Some students will probably state that sand is a liquid because it flows. You can acknowledge their excellent thinking, and perhaps explain that to come to agreement on an issue such as this, scientists have to decide among themselves what they mean when they say *sand*. By convention, then, they agree that they mean (in this context) to focus on each grain, in which case it is clear that sand is a solid.

Exotic States of Matter — Beyond Solids, Liquids, and Gases

Matter such as Glook, toothpaste, and shaving cream do not fit easily into the categories of solids, liquids, or gases. Scientists understand these materials as combinations of solids, liquids, and gases that interact in interesting ways. Listed below are a few forms of matter that might be considered genuinely different from solids, liquids, and gases. We do not recommend discussing these forms with your students; this is for your information only.

Plasma

A **plasma** is an ionized (charged) gas. It therefore has many of the same properties as other gases. It flows and it expands to fill empty spaces. What makes plasma different from ordinary gas is that its particles are so energetic that the electrons separate from the molecules. This forms a mixture of negative electron gas and positive ion gas. Unlike ordinary gases, plasma can conduct an electric current. Plasma is less exotic than you might think. Fluorescent light bulbs, neon signs, lightning bolts, and some television screens have plasma in them. Stars are mostly made of plasma, which means that most of the *visible* Universe is plasma.

Bose-Einstein Condensate

While plasma is very hot, a **Bose-Einstein condensate** is very cold. In certain materials, when they are cold enough, atoms become identical to each other and can essentially occupy the same place at the same time. In 1938 this effect was observed in a certain kind of liquid helium when it was cooled to about two degrees above absolute zero. The liquefied helium turned into a "superfluid" that flows with absolutely no friction. A true Bose-Einstein condensate forms from a gas. This was first achieved in 1995 when a few hundred atoms of rubidium vapor were cooled to a tiny fraction of a degree above absolute zero. This work was awarded the Nobel Prize in Physics in 2001. There are still many unanswered questions about the nature of this bizarre state of matter. Bose-Einstein condensates are observed only in specialized laboratories.

Free Particles

The categories of solids, liquids, and gases all are based on how the particles in a substance relate to each other. Yet the Universe is full of particles that fly through space on their own. In space, in high-energy physics laboratories, and even in television picture tubes, there are unbound bits of matter. Since they rarely interact with each other they are usually just considered *free particles.*

Dark Matter

Most of the light from distant galaxies is from hot plasma in stars. That is the matter we can see. Observations of the motions of galaxies give evidence that galaxies are affected by matter that has never been observed. This is what has been called **dark matter.** Since this matter cannot be seen, scientists are not sure what it is exactly, or what form it is in, but there is evidence that most (!) of the Universe is composed of dark matter. Perhaps some dark matter is in forms that are already known but not bright enough to detect. The possibility exists that this mysterious matter is in some completely different form.

A Note on Phase Change

While this unit does not delve into phase change, it sets the stage for it, and students may ask questions about it, so this brief background may be helpful. Substances—such as water, gold, oxygen, and so on—can be solid, liquid, or gas, depending on the temperature and air pressure around them. When enough heat is added to a solid it will change to liquid phase, in the process called **melting.** The temperature at which this happens is called the *melting point.* If enough heat is added to the liquid, it will change to gas phase, in the process called **evaporation.** The particles (atoms or molecules) in a liquid need to gain enough energy to change to gas phase. Evaporation only happens at first on the surface of a liquid. Evaporation can happen at even low temperatures, although it's much faster at higher temperatures. When fast-moving liquid particles get enough energy to escape as gases, the particles left behind have less energy, and the remaining liquid is cooler. This is called **evaporative cooling**, and it's the reason a person feels colder as water evaporates off their skin. The rate of evaporation is also affected by the humidity and movement of air around the liquid. If a liquid is heated enough, it will change to gas phase *below* the surface too. This is called **boiling**, and the lowest temperature at which it happens is called the *boiling point.*

If enough heat is taken away from a gas it will change to liquid phase, which is called **condensation.** The tiny droplets of water that collect on the outside of a cold can, or the dew on grass in the morning, are both examples of the condensation of water vapor in the air. If enough heat is taken away from a liquid it will change to solid phase, which is called **freezing.** The temperature at which this happens is called the *freezing point.* The melting and freezing point of a pure substance is usually the same temperature. For example, water freezes at 32°F (0°C) and below; ice melts at 32°F (0°C) and above.

Sometimes when heat is added, a solid may change directly from solid phase to gas phase, without going through a liquid phase. This is called **sublimation.** Dry ice is a substance that sublimates at the pressure of air in Earth's atmosphere, so it doesn't become liquid before changing to gas phase. The opposite is sometimes true when heat is taken away from a gas and it changes directly to solid phase. This is called **deposition.** Frost and snow are both formed through deposition when water vapor changes directly into solid ice. **No matter what phase a substance is in, it is still the same substance.** Water is water, whether it is solid ice, liquid, or water vapor. It has some different characteristics at each phase, but it can always be changed to another phase by adding or removing heat.

A more accurate demonstration that air has mass can be done by measuring the increased weight when pressurizing a readily available 1-liter PET bottle, the type used for carbonated soda. A PET bottle can be safely pressurized to over 100 pounds per square inch (psi), although this demonstration works satisfactorily with pressures around 30 psi. A bicycle air valve or Schrader valve (available at auto supply stores) is inserted in a hole drilled in the screw-on cap. A waterproof glue such as polyurethane construction adhesive found in home improvement stores (PL Premium) is used to seal the outside of the valve to the cap. Once dry and assembled, weigh the bottle with the most sensitive scale you have. Using a bicycle pump, pump air into the bottle, then weigh it again. The bottle weighs more because there is more air in the bottle, and air has mass. With two similar bottles and a pan balance you have a nice, visual display that air has mass. To make sure the two bottles weigh exactly the same at the start, you may need to add a bit of clay to the outside of whichever bottle is lighter. You can set up a similar demonstration using only one altered bottle if you add enough clay to the other bottle to make sure they weigh the same before adding air to the bottle with the valve. The bottles can be passed around for students to feel the wall stiffness in the bottle that is pressurized. Of course when the cap is loosened, the excess pressure is vented and the bottle returns to its former mass, equalizing the balance. Using bottles for these demonstrations works better than using balloons because the volume of the water bottle does not change with pressurization like that of a balloon does.

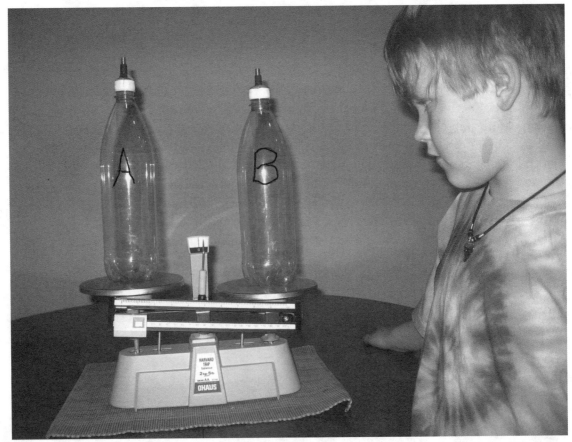

TEACHER'S OUTLINE

Activity 1: Solids and Liquids

■ Getting Ready: Before Day of Activity
1. Decide if you need to pre-teach sorting.
2. Find space for class display.
3. Prepare bags of solid and liquid items.

■ Getting Ready: On Day of Activity
1. Write on a sentence strips, but don't post yet:
   — Hold their shape and do not turn into a puddle.
   — Take the shape of their container.
   — Stay flat on top, unless moving.
   — Don't hold their shape, and do make a puddle.
2. Make large Solids and Liquids signs.
3. Gather items for whole-class demonstrations.
4. Copy two student sheets and plan groups of four.

■ Observing Collections of Objects
1. What's in the Universe?
2. Free exploration of objects.
3. Each group will get a bag of objects.
4. Divide the class into groups of four.

■ Sorting the Objects
1. After free exploration, regain attention of class.
2. Introduce sorting activity.
3. Have students begin.

■ Introduce the Secret Sort
1. Gather class away from materials.
2. Briefly discuss their sorts.
3. Explain Secret Sort game.
4. Set cards labeled #1 and #2 on carpet a foot or two apart.
5. Say hard part of game is need to be silent.

■ Play the Secret Sort game
1. Dramatize your sorting process; students make predictions silently as in guide.
2. Ask, "What is the same about everything in Group #1?" Before you reveal secret rule, call on a few students and test their ideas.
3. Reveal "secret" of Group #1 and place Solids sign in front of it.

■ Define Solids
1. Ask for their ideas about solids.
2. Scientists define a solid as *something that holds its shape.*
3. Set statement "Hold their shape and do not make a puddle" below the "solids" sign.

■ Define Liquids
1. Ask, "What is the same about everything in Group #2?" Every object in Group #2 is a liquid.
2. Place Liquids sign in front of Group #2. Ask, "What makes something a liquid?"
3. Scientists define a liquid as *something that does not hold its shape.*
4. Set the statement "Don't hold their shape and DO make a puddle" below the liquids sign.
5. Add two more true statements about liquids: "Take the shape of their container" and "Stay flat on top, unless moving." Challenge students to come up with more true statements about solids and liquids.

■ Re-Sorting Objects and Recording in Journals
1. Students re-sort objects into solids and liquids, using senses for evidence.
2. What is evidence? If we can see or feel that something holds its shape, that's evidence that it is a solid. Evidence is information from our senses.
3. Explain and hand out student sheets and help students set up journals.
4. Groups return to bags of objects. Assemble signs and strips on display as in guide.
5. Circulate to assist students having trouble with categories and definitions.
6. Clean up. Have students put objects back in bags; collect bags and journals.

Activity 2: Collecting Solids and Liquids

■ Getting Ready
1. Before day of activity, decide whether to have adult volunteers.
2. On day of activity, set out plastic bags.
3. Have sentence strips, marker, and pushpins near display.
4. Write procedure steps in large letters on board or chart paper.
5. Set up 10 learning stations, as detailed in guide.

■ Reviewing Solids
1. Review what a solid is, and definition of evidence.
2. Ask students to think of a solid and raise hands.

3. Have them all share as described in guide.

4. At count of three, every student points to a solid in the room.

5. To reinforce definition, describe where a few students are pointing.

■ Reviewing Liquids

1. Review definition of liquids.

2. Tell students to think of a liquid, doesn't need to be in room. Anything they drink is a liquid, but there are many liquids they can't drink—like poison.

3. As before, go around circle, calling on all students.

4. Reinforce definition of a liquid.

■ Introducing the Solids and Liquids Learning Stations

1. They will go to stations with partner, collect samples, and add to class display.

2. Model five-step procedure. A *procedure* is all steps used to do task.

3. Reviewing procedure by modeling a "wrong way."

4. Assign pairs to first stations.

■ Conducting the Learning Station Activity

1. Circulate, helping out as needed.

2. Be sure both partners record in journals before going to a new station.

3. Give five-minute signal. Ask a few pairs to tidy stations; rest to discussion area.

■ Discussing the Collections on the Class Display

1. Gather students to see display.

2. Discuss classifications.

3. Discuss chalk dust and crushed cereal.

■ Adding to the Definitions of Solids and Liquids

1. You'd like their help in adding to definitions of solids and liquids.

2. Ask, "Is there anything else that is true of all solids or all liquids?"

3. Post new definitions.

4. As possible, illustrate sentence strips with drawings.

5. Have students add new definitions to journals.

Activity 3: Challenging Substances

■ Getting Ready

1. Make one copy per student of Solid or Liquid? sheet.

2. Have available: solid and liquid items, baking soda in cup, journals.

■ Set Up the Learning Stations

1. Set up shaving cream, toothpaste, sand, Glook stations.
2. Introduce solid and liquid hand signals as in guide.

■ Introducing Challenging Substances

1. Draw attention to display.
2. Are shaving cream, toothpaste, Glook, and sand solids or liquids?
3. Go to stations in any order—but do not collect samples. Try to figure out if each is solid or liquid and record decision.
4. Only four different stations; there are duplicates of each.

■ Introducing the Procedure

1. Show how to record on Solid or Liquid? sheets.
2. Demonstrate steps for each station, as detailed in guide.
3. As students watch, go to the stations and show rules to control mess.

■ Conducting the Learning Station Activity

1. Pass out Solid or Liquid? sheets
2. Assign pairs to first stations, and let them begin.
3. Circulate; remind students to record their reasoning.
4. Give five-minute warning.

■ Discussing Glook, Toothpaste, Shaving Cream

1. Discuss Glook.
2. Discuss toothpaste and shaving cream.
3. Other scientists agree these are tricky to sort into solids or liquids.
4. Reveal that scientists classify these three as *colloids*.

■ Discussing Sand

1. Ask for hand signals. Hold up a sand sample; sand *does* fit into one of the categories.
2. Ask students to share evidence that sand is a solid or liquid.
3. Guide them to understanding that each grain is a solid.
4. Remind them of chalk. "Was it still a solid when broken up into small pieces?"
5. Piles of tiny solids are still solid; each piece holds its shape.

■ Explaining to Someone Else Why Powders are Solids

1. Hold up cup of baking soda—like sand, only pieces are smaller.
2. Pretend you think baking soda is a liquid. Ask students to explain what may be incorrect about statements in guide.
3. Students add Solid or Liquid? sheet to journals.
4. In journals—write a letter to someone who doesn't think sand is a solid.

Activity 4: What's the Matter?

■ Getting Ready

1. Attach bag to spring scale for each pair, as specified in guide.
2. Make two large signs—MATTER and NOT MATTER.
3. Make more signs: have mass; take up space; air; people; do not have mass; do not take up space.
4. Gather scale and items to demonstrate how to use.
5. Inflate balloon with carbon dioxide, as detailed in guide.
6. Set out objects for Matter testing stations.
7. Make copy of What's the Matter? student sheet for each student.

Introducing Matter

■ Testing If Objects Take Up Space

1. Introduce matter. Some things in Universe belong in a group called *matter*.
2. To be matter, must take up space. Put "take up space" sign near Matter sign and "do not take up space" near the "not matter" sign.
3. "Does a pencil take up space?" Show how you can put your fingers together when pencil not between them.
4. Pinch pencil between fingers. When you try to put fingers together, pencil blocks; evidence that pencil is taking up space.
5. Students use this "finger test" to see if objects take up space.

■ Testing If Objects Have Mass

1. To be matter, something has to have *mass* (it is made out of "stuff.")
2. Put the "have mass" sign near Matter sign and the "do not have mass" near the "not matter" sign.
3. Ask how they could test a pencil to see if it has mass. Accept all ideas.
4. Gravity is a pull between Earth and the pencil; gravity pulls pencil toward ground. If pencil did not have mass, gravity would not pull on it, and it wouldn't feel heavy.
5. Test for mass using a scale. If object weighs something, that's evidence it has mass.

■ Demonstrating How to Use the Scales

1. Hold up a spring scale. Ask what will happen if you put pencil into bag.
2. Set pencil in bag; the more scale moves, the more mass (and weight) an object has.
3. Test heavier object. Have students predict what will happen. Try it.
4. Use object that is too heavy to be measured by the scale. It pulls down on the spring as far as it can go. Ask if the object has mass.
5. Caution students to treat the spring scales carefully.

■ Circulating to Stations

1. Ask, "How do you know the pencil is matter?" [It takes up space and has mass.]
2. Show how to record on What's the Matter? sheet.
3. If testing something not on list, use a blank space, do tests, answer questions.
4. Pass out materials and begin. They can go to stations in any order.

■ Discussing Results

1. Keep What's the Matter? sheets and gather for discussion.
2. Did anything surprise them? Ask, "What is one thing you tested that is matter?" For each response, ask how they know. [It has mass and takes up space.]
3. Ask if they found any solids that are not matter? Any liquids?
4. Discuss the water in the balloon.
5. *All* solids and liquids are matter. Move Matter sign above Solid and Liquid signs.

■ Testing "Air"

1. Is air matter? Students hold thumbs up or down.
2. Show deflated balloon. Put on scale and record grams. Hold up CO_2-filled balloon. Balloon itself has same amount of mass as the other, but this one has air in it.
3. Put CO_2 balloon on scale. Ask, "Does the balloon with air in it have more mass than the empty balloon? Why?"
4. Students use fingers to test if air in room takes up space. They may say no because they can easily put their fingers together.
5. Explain that air is moving out of the way.
6. Students do finger test on filled balloon. Ask, "Can you feel something between your fingers that takes up space?"
7. Is air matter? [Yes, has mass and takes up space.] Put the balloon and "air" sign under Matter sign.

■ What's *Not* Matter?

1. Can they think of things that are not matter—that don't have mass or take up space. Place the "Not Matter" sign to the side of display.
2. Give some examples as needed.
3. Ask students to think of other things that are not matter for next session. Have them write them on a sheet of paper, title it Not Matter, and add it to their journals.

■ Are People Matter?

1. Ask volunteer to stand in front of class. Are people matter?
2. Put your hands on either side of volunteer's head. You can't put your hands together because something between them takes up space.

3. Do people have mass? Lift student or pretend to try, and tell them that yes, you felt weight. Confirm that the evidence shows that people do have mass.

4. Ask again if people are matter. [Yes, they have mass and take up space.] Place "people" sign under Matter sign.

Activity 5: Gases

Session 1: How Do We Know Gases Exist?

■ Getting Ready

1. Copy large Gases sign.

2. Write sentence strip sign "spread in the air."

3. Have peppermint extract but out of sight.

4. Fill clear container or cup partially with water.

■ Introducing Gases

1. Review any examples of "not matter" posted. Elicit more.

2. Hold up container partially filled with water; students show hand signals for solid (a fist) or liquid (slowly waving hand).

3. What else is in the container? Tell them air is a mixture of *gases.*

■ Thinking of Evidence that Gases Exist

1. Pretend you don't think gases exist.

2. Challenge students to provide evidence that gases/air exist or are real.

3. List their suggestions and acknowledge that gases exist.

■ Gases Are a Third Kind of Matter

1. Ask, "Is gas matter?" Remind class how by squeezing balloon with fingers, a person could feel that air/gas takes up space. Gas is matter because it has mass and takes up space.

2. Add Gases sign to display. Gases are a third kind of matter.

3. Give hand signal for gases: wiggling fingers.

■ Peppermint Demonstration: Gases Can Spread Out and Move Around

1. Our noses are gas-detectors.

2. Teacher will release a pleasant-smelling gas. They wait with eyes closed, raise hands if they smell it.

3. When eyes are closed, open peppermint extract and pour on a dish.

4. Who smells peppermint? Show liquid peppermint in dish.

5. How did the peppermint smell get to noses? The smell is evidence that the gas came from peppermint on plate and moved through air.

6. Gases move and spread around more easily than solids or liquids do.

7. Post "spread in the air" under Gases sign.

Session 2: Investigating Gases

■ Getting Ready: Before the Day of the Activity

1. Decide how many stations to set up and how to arrange them.
2. Arrange for two adult volunteers at *Making Gas* stations.
3. Have a helium balloon brought in.

■ Getting Ready: On the Day of the Activity

1. Set up the stations as detailed in guide, and a labeling station.
2. Have a few materials from previous sessions, including container half full of water.
3. Make photocopies.

■ Gas Stations

1. Gases are a third kind of matter.
2. Collect gases only at Stations #1 and #2.
3. Point out labels, pens, and pushpins; tell how to pin sample on display.
4. Explain procedure at each station.
5. Assign pairs to their first station.

■ Discussing Gas Stations

1. Gather for class discussion. Ask students to think of evidence from stations that gas exists. Discuss Station #1 last. For each station, ask, "What is your evidence that gas exists, even though you can't see it?"
2. Use Station #1 to emphasize evidence that gas is matter.
3. Does gas take up space? Try to put fingers together around gas-filled balloons. Tell students that all gases take up space.
4. Ask, "Which of the gases had the most mass?" "Least mass?"
5. Helium does have mass, but less mass than air.
6. Show picture of helium balloon in vacuum; explain why it sinks to bottom.
7. Ask again—is gas matter?

■ More Challenging Substances

1. Show hand signals for solid and liquid. Remind them of signal for gas.
2. Hold up a few diverse items and have them signal what they are.
3. They could show all three signals at once.

■ Classifying People

1. People are matter.

2. Which kind(s) of matter are we? Ask them to use their hand signals.

3. Share ideas and review definitions.

4. People are made of all three kinds of matter, as are all living things.

5. Scientists study matter in the Universe.

■ Gases Student Sheet

1. Introduce student sheet. Ask for ideas on how to draw gases.

2. Students share examples of gases at stations. Show how to draw and label gases.

3. Pass out sheets and have students begin.

ASSESSMENT SUGGESTIONS

Anticipated Student Outcomes

1. Students are able to compare and sort objects according to their observable properties.

2. Students can define the properties of and discuss the differences between solids, liquids, and gases.

3. Students are able to describe the characteristics of certain "challenging substances" that are difficult to classify as solid or liquid.

4. Students understand that gases are matter because gases take up space and have mass.

5. Students deepen their understanding of matter as something that takes up space and has mass, and are able to explain why all solids, liquids, and gases are matter.

6. Students demonstrate increased understanding of what scientific evidence is, and improve in their ability to support their explanations with evidence.

Embedded Assessment Activities

Sorting Solids and Liquids. In Activity 1 students engage in a series of sorting activities. Your observations in class and review of the student sheets should provide an initial sense of how well students are grasping the distinction between solids and liquids. (Addresses Outcomes 1 and 2)

Sand is a Solid Letter. At the end of Activity 3, under "Explaining to Someone Else Why Powders are Solids," students are to write a convincing letter in their journals to someone who doesn't think sand is a solid. This letter can provide information on student understanding of substances such as sand, baking soda, or other powdery solids. The letter also can shed light on student understanding of evidence through their use of evidence in their arguments. Your observations of student responses when you pretend that you think baking soda is a liquid can also be helpful in assessing student understanding. (Outcomes 2, 3)

Challenging Substances. The **Solid or Liquid?** student sheet, which students complete in Activity 3, can provide very useful information on student understanding of what has been learned in the unit up to this point. Students are asked whether they think shaving cream, toothpaste, Glook, and sand are solid, liquid, or "not sure." They are asked to explain their reasoning for each substance. (Outcomes 2, 3, 6)

Gases Exist. In Activity 5, Session 1, "Thinking of Evidence to 'Prove' Gases Exist," the teacher pretends to be someone who thinks gases do not exist. Students are challenged to "prove" that gases/air exist and are real. Student responses and your listing of their responses can provide some insight into the class level of understanding about gases. (Outcome 4)

Gases Student Sheet. In Activity 5, Session 2, the **Gases** student sheet provides further information on student thinking about and understanding of gases. It's also an opportunity for those students who may be more inclined toward visual representations to portray their understanding through drawing. (Outcomes 4, 5)

Additional Assessment Ideas

Special Assessment. The "Solid, Liquid, or Gas?" and "Matter or Not Matter?" three-page questionnaire can serve as an excellent assessment for the major concepts in the unit. It is on pages 110–112. While it has been designed to provide reliable information about student learning, it, like any other assessment tool, should not be seen in isolation, or as a final measure of what students have learned, but considered along with other assessments suggested here as well as with your own observations and evaluations of student progress. (Outcomes 2, 4, 5, 6)

Solid and Liquid Writing Assignment or "Quick Writes." In general, asking students to describe their thinking in writing (and/or drawing) at different stages in the unit can be very helpful. One of the "Going Further" suggestions for Activity 1 includes a writing assignment in which students pick one solid from the objects they sorted, draw it on a separate piece of paper, and describe why they classified it as a solid, then do the same with a liquid. A "Going Further" for Activity 2 suggests a "Quick Write" in which students describe their evidence for whether they think chalk (or whatever substance you used at the "crushing" station) is a solid or a liquid. At these points in the unit, these assignments can help you assess student understanding. (Outcome 1; other outcomes depending on assignment).

Solid, Liquid, or Gas?

Juice	Circle one:
	solid liquid gas
Explain your answer.	

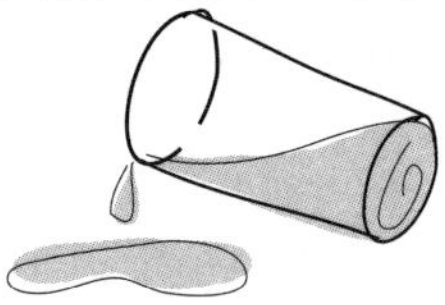

_____

_____

_____

_____

_____

_____

Air	Circle one:
	solid liquid gas
Explain your answer.	

_____

_____

_____

_____

_____

_____

Salt

Circle one:

solid   liquid   gas

Explain your answer.

_____

_____

_____

_____

Spoon

Circle one:

solid   liquid   gas

Explain your answer.

_____

_____

_____

_____

An Orange

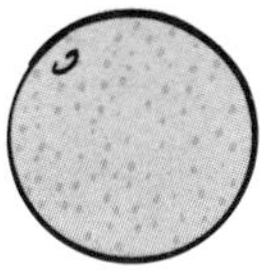

Circle one:

solid   liquid   gas

Explain your answer.

_____

_____

_____

_____

Matter or Not Matter?

Are these things matter or not matter? Explain how you know.

juice
Matter or Not Matter

Explain your answer.

_____

_____

_____

feelings
Matter or Not Matter

Explain your answer.

_____

_____

_____

air
Matter or Not Matter

Explain your answer.

_____

_____

_____

RESOURCES AND LITERATURE CONNECTIONS

Sources

These activities generally utilize readily available materials. Helium balloons (Activity 5, Session 2) are of course available at party stores and many large supermarkets. However, there is one set of needed equipment—the spring scales (one for the teacher and one for every pair of students)—that need to be obtained (or borrowed) for use in Activity 4. The GEMS Kit for this unit from Carolina Biological includes one calibrated spring scale for the teacher and 16 un-calibrated ones for students. These un-calibrated scales are less expensive and fine for use in these activities. Many science suppliers and kit producers carry various brands of spring scales, including less expensive un-calibrated items. On their websites, search under "spring scales." Here are some sources.

Carolina Biological Supply Company
2700 York Road
Burlington, NC 27215-3398
800 227 1150
Fax: 800-222-7112
www.carolina.com/GEMS
gemskits@carolina.com

Delta Education
80 Northwest Blvd.
P.O. Box 3000
Nashua, NH 03061-3000
800-258-1302
Fax: 800-282-9560
http://www.delta-education.com

NASCO
901 Janesville Avenue, P.O. Box 901
Fort Atkinson, WI 53538-0901
4825 Stoddard Road, P.O. Box 3837
Modesto, CA 95352-3837
1-800-558-9595
Fax: 920-563-8296
http://www.enasco.com/

Science Lab
14025 Smith Rd.
Houston, Texas 77396
1-800-901-7247
Fax: 281-441-4409
http://www.sciencelab.com/

Science Kit & Boreal Laboratories
777 E. Park Drive
PO Box 5003
Tonawanda, NY 14150
800-828-7777
Fax: 800-828-3299
http://www.sciencekit.com/

Related Curriculum Material

The Full Option Science System (FOSS), Solids and Liquids Module. This module for first and second graders provides a variety of experiences with solids and liquids to heighten students' awareness of the physical world.

Insights: An Inquiry-Based Elementary School Science Curriculum, Liquids Module. In this second/third grade module, students explore the unique characteristics of liquids, compare different liquids, and explore how solids and liquids interact with each other.

The Science and Technology for Children (STC), Solids and Liquids Unit. In this Kindergarten/first grade unit, students investigate the similarities and differences in a variety of common solids and liquids.

Nonfiction for Students

Change It!: Solids, Liquids, Gases and You
(Primary Physical Science).
By Adrienne Mason
Kids Can Press, 2006

A Chilling Story: How Things Cool Down
by Eve Stwertka
Julian Messner, 1991

Experiments With Solids, Liquids, and Gases (True Books: Science Experiments)
by Salvatore Tocci (Author)
Children's Press, 2002

Matter and Materials (Hands-on Science series)
by Sarah Angliss
Kingfisher Books, Houghton-Mifflin, 2001

The Microscope
by Maxine Kumin
Harper and Row, 1984

The Popcorn Book
by Tomie dePaola
Holiday House, 1984

The Quicksand Book
by Tomie dePaola
Holiday House, 1977

The Slimy Book
by Babette Cole
Red Fox, 2003

Solids, Liquids, and Gases
(Rookie Read-About Science series)
by Ginger Garrett
Children's Press, 2005

Solids, Liquids and Gases
(Starting with Science series)
by Ontario Science Centre
Kids Can Press, 1995

Solid, Liquid, or Gas? (It's Science series)
by Sally Hewitt
Children's Press, 1998

What Is Matter?
(Rookie Read-About Science series)
by Don L. Curry
Children's Press, 2005

What Is the World Made Of?
All About Solids, Liquids, and Gases (Let's-Read-and-Find-Out Science, Stage 2)
by Kathleen Weidner Zoehfeld and Paul Meisel
HarperTrophy, 1998

Fiction for Students

Bartholomew and the Oobleck
by Dr. Seuss
Random House, 1949

Everybody Needs a Rock
by Byrd Baylor
Aladdin, 1985

Hot-Air Henry (Reading Rainbow Books)
by Mary Calhoun
HarperTrophy, 1984

Water's Way
by Lisa Westberg Peters
Arcade Publishers, 1991

Internet Sites

http://www.chem4kids.com/files/matter_intro.html

http://www.bbc.co.uk/schools/scienceclips/ages/8_9/solid_liquids.shtml

http://lhsfoss.org/fossweb/teachers/audio/audio_books/SOLIDS_AND_LIQUIDS.MP3

http://www.strangematterexhibit.com/

REVIEWERS

We warmly thank the following educators, who reviewed, tested, or coordinated the trial tests in manuscript or draft form. Their critical comments and recommendations, based on classroom presentation of these activities nationwide, contributed significantly to this GEMS publication. (The participation of these educators in the review process does not necessarily imply endorsement of the GEMS program or responsibility for statements or views expressed.) Classroom testing is a recognized and invaluable hallmark of GEMS curriculum development; feedback is carefully recorded and integrated as appropriate into the publications. WE THANK THEM ALL! ■

CALIFORNIA

Mills College Children's School, Oakland
Mari Litsky
Michelle Quraishi

Webster School, Oakland
Harriet Frank
Carol Murayama
Pali Ouye

Dingeman Elementary School, San Diego
Joe Anthony
Charlyne Barad
Michelle Becker
Leah Frost
Suzanne Hagan
Ann Webb

St. Paul School, San Pablo
Liz Gray
Michelle Hathorne
Kathleen Kraft

Toyon Elementary, Shasta Lake
Kathleen Evanhoe
Ramona Guerra
Colleen Lytle
Joanne Scrima

Junction Elementary School District, Somes Bar
Chris Magarian

MINNESOTA

East Bethel Community School, Cedar
Pamela Beecham
Ann Hove
Judy P. Nelson

NEW YORK

The Browning School, New York City
Janet Lopez

OREGON

Kennedy Elementary School, Medford
Teena Staller
Patty Munson
Patty Drake
Debbie Clark
Terry James

Abraham Lincoln Elementary, Medford
Kathy Staller

Willamette Primary School, West Linn
Debbie Nicolai
Amy Thibault
Carlene Kesterson
David Pryor

SOUTH CAROLINA

Pauline-Glenn Springs Elementary School, Pauline
Spiller Gregory
Garrison Hall
Cindy Hendrix
Lorraine Walker
Ellen Wetmore

TEXAS

Gilbert Elementary, Irving
Brant Darnell
Shanon Horak
Carol Johnson
Ruby Laney

Irving Independent School District, Irving
Jacque Garcia

Nichols Sawmill Elementary, Magnolia
Lynet Fox
Stefani Kulhanek
Lisa Lowery
Jennifer McConnelee

WISCONSIN

St. John Lutheran School, Wauwatosa
David L. Alleheilign
Marlis Kramer
Lynn Schmidt

SOLIDS

SATURN

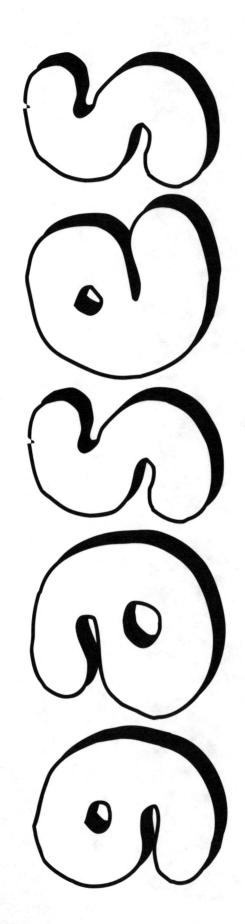

Get Connected – Free!

Get the *GEMS eNews*,

our free e-mail newsletter filled with...
- *updates* on GEMS activities and publications
- **information** about workshops and leadership training
- **announcements** of new publications and resources

And...receive our print catalog!

*Be part of a growing national network of people who are committed to excellence in math and science education. Stay connected with the **GEMS** eNews. Simply return the attached postage-paid card.*

For more information about GEMS call (510) 642-7771 or write to us at GEMS, Lawrence Hall of Science, University of California, Berkeley, CA 94720-5200, or gems@berkeley.edu

Please visit our web site at lhsgems.org

GEMS activities are effective and easy to use. They engage students in standards-based, inquiry-driven math and science explorations, while introducing key principles and concepts.

More than 80 GEMS Teacher's Guides and Handbooks have been developed at the Lawrence Hall of Science — the public science center at the University of California at Berkeley — and tested in thousands of classrooms nationwide. There are many more to come — along with GEMS Workshops and GEMS Centers and Network Sites springing up across the nation and world to provide support, training, and resources for you and your colleagues!

Yes!

Sign me up for a free subscription to the

GEMS eNews

filled with ideas, information, and strategies that lead to Great Explorations in Math and Science!

Name_____

Address_____

City_____ State_____ Zip_____

e-mail address:_____

How did you find out about GEMS? (Check all that apply.)
❑ word of mouth ❑ conference ❑ ad ❑ workshop ❑ other: _____
❑ Please send me a free catalog of GEMS materials.

GEMS
Lawrence Hall of Science
University of California
Berkeley, CA 94720-5200
(510) 642-7771

Ideas◄
Suggestions◄
Resources◄

that lead to Great Explorations
in Math and Science!

Sign up
now for a
free subscription
to the *GEMS*
eNews!

01 LAWRENCE HALL OF SCIENCE # 5200

-61571-25775-62-X

Get Connect